To my very good
Peter & Julia, with very
best wishes from the
author!
John

From Servant to Queen

A Journey through Victorian Mathematics

With a few exceptions, pure mathematics in Britain at the beginning of the nineteenth century was a recreation for amateurs. Drawing on primary sources, John Heard provides an engaging account of the process by which it rose to become an academic discipline of repute which by the First World War was led by G H Hardy, and supported by the internationally respected London Mathematical Society. In chronicling that rise, this book describes key contributions and the social environment in which mathematicians operated, uses contemporary commentary where appropriate, and provides full references to help any researchers who want to dig deeper into the original sources. No mathematical knowledge is required, and readers with a wide range of interests and backgrounds will find much to enjoy in this unique insight into the world of Victorian mathematics and science.

John Heard is an independent researcher with a doctorate in the history of science from Imperial College London. He is also a member of the British Society for the History of Science, the British Society for the History of Mathematics, and a Fellow of the Institute of Chartered Accountants in England and Wales.

Arthur Cayley, painted by Lowes Cato Dickinson in 1872; a satirical verse by James Clerk Maxwell concludes with the lines:

> ... by Dickinson depicted
> In two dimensions, we the form may trace
> Of him whose mind, too large for vulgar space,
> In n dimensions flourished unrestricted.

From Servant to Queen
A Journey through Victorian
Mathematics

John Heard

CAMBRIDGE
UNIVERSITY PRESS

CAMBRIDGE
UNIVERSITY PRESS

University Printing House, Cambridge CB2 8BS, United Kingdom

One Liberty Plaza, 20th Floor, New York, NY 10006, USA

477 Williamstown Road, Port Melbourne, VIC 3207, Australia

314–321, 3rd Floor, Plot 3, Splendor Forum, Jasola District Centre, New Delhi – 110025, India

79 Anson Road, #06–04/06, Singapore 079906

Cambridge University Press is part of the University of Cambridge.

It furthers the University's mission by disseminating knowledge in the pursuit of education, learning, and research at the highest international levels of excellence.

www.cambridge.org
Information on this title: www.cambridge.org/9781107124134
DOI: 10.1017/9781316415726

© John Heard 2019

This publication is in copyright. Subject to statutory exception and to the provisions of relevant collective licensing agreements, no reproduction of any part may take place without the written permission of Cambridge University Press.

First published 2019

Printed in the United Kingdom by TJ International Ltd. Padstow Cornwall

A catalogue record for this publication is available from the British Library.

Library of Congress Cataloging-in-Publication Data
NAMES: Heard, John (John Michael), author.
TITLE: From Servant to Queen : A Journey through Victorian Mathematics / John Heard.
DESCRIPTION: Cambridge ; New York, NY : Cambridge University Press, 2019. | Includes bibliographical references and index.
IDENTIFIERS: LCCN 2018042537 | ISBN 9781107124134
SUBJECTS: LCSH: Hardy, G. H. (Godfrey Harold), 1877–1947. | Mathematicians – Great Britain – Biography. | Mathematics – History – 19th century. | Mathematics – History – 20th century. | Aesthetics. | Mathematics in nature.
CLASSIFICATION: LCC QA29.H23 H43 2019 | DDC 510.92/241–dc23
LC record available at https://lccn.loc.gov/2018042537

ISBN 978-1-107-12413-4 Hardback

Cambridge University Press has no responsibility for the persistence or accuracy of URLs for external or third-party internet websites referred to in this publication and does not guarantee that any content on such websites is, or will remain, accurate or appropriate.

This book is dedicated to the memory of my dear parents, Nancy and Percy.

CONTENTS

	List of Figures	page viii
	Acknowledgements	ix
	Notes for the Reader	x
1	Setting the Scene	1
2	The Legacy of Newton	11
3	The London Mathematical Society	50
4	The Pure Mathematician as Hero	111
5	Mathematicians in an Aethereal World	145
6	Apologias for Pure Mathematicians	190
7	Embracing Beauty	214
	Epilogue	243
	Select Bibliography	247
	Index	257

FIGURES

3.1	Invitation to the first meeting of the 'University College Mathematical Society'	page 58
3.2	Original design by Augustus De Morgan of a badge for the London Mathematical Society	62
3.3	Construction diagram from the first proposition in Euclid's *Elements*	63
3.4	Total number of ordinary members of the London Mathematical Society at each position in the table of wranglers	73
4.1	James Whitbread Lee Glaisher	116
4.2	Henry John Stephen Smith	129
5.1	Chart showing references to 'pure mathematician(s)'	146
5.2	The controversial 'parallel postulate' from the *Elements of Euclid*	167
6.1	Sylvester's characterisation of mathematical discovery	207
7.1	Cartoon from *Punch* by George du Maurier	219
7.2	Lois Cayley: clearly a mathematician	230
7.3	The existence proof: Hardy on the left, Littlewood on the right	238

ACKNOWLEDGEMENTS

This book could not have been written without access to many libraries and archives, and I extend my sincere thanks to the librarians and archivists who, over the years, have allowed me to consult the materials in their care. Some of these materials are quoted or reproduced here, and in this respect I am pleased to make the following acknowledgements: letters from Henry Baker, by permission of the Master and Fellows of St John's College, Cambridge; the portrait of Arthur Cayley, the photograph of G H Hardy and J E Littlewood, a letter from Harald Cramér, and an obituary speech by Alexander Ostrowski, by permission of the Master and Fellows of Trinity College, Cambridge; letters from Andrew Russell Forsyth, by permission of the British Library; a letter from G H Hardy, by permission of the Syndics of Cambridge University Library; the journal of Thomas Archer Hirst, by courtesy of the Royal Institution of Great Britain; minute books of the London Mathematical Society, by courtesy of the Society; portrait of J W L Glaisher, by permission of the Royal Astronomical Society; 'My First Gorilla', by permission of the Zoological Society of London; a sketch of a badge for the London Mathematical Society, a letter to Thomas Archer Hirst, and letters from Sophia De Morgan, by permission of University College (London) Library, Special Collections.

I am grateful to June Barrow-Green for an initial suggestion that led to the writing of this book, and to her and Tony Crilly for their valuable comments on a draft of it; of course, the responsibility for any errors or omissions is mine alone. Finally, my partner Stella Pilavaki has provided constant and loving support and encouragement throughout the authorial process.

NOTES FOR THE READER

Two forms of referencing are used in this book. An alphabetic superscript refers to a short note at the bottom of the page; such notes give additional information or explanations that do not fit easily in the main text. Much more numerous are numerical superscripts, which refer to the 'Notes and References' section at the end of each chapter. The information here concerns sources, and nothing else; so if the reader is not concerned about these, then *all numerical superscript references can safely be ignored*. In each 'Notes and References' section, full bibliographic details are given for the first mention of a publication, with an abbreviated reference being given for subsequent mentions.

For convenience and conciseness, the term 'Britain' is used to signify Great Britain and the island of Ireland; while 'the Continent' signifies that part of Europe on the other side of the English Channel.

In the index, if the title of a book or journal begins with a definite or indefinite article, then the article is omitted; thus *A Budget of Paradoxes* in indexed under 'B', and 'The Purloined Letter' under 'P'.

In quotations, any italicised stress is in the original.

1 SETTING THE SCENE

It is possible that the life of a mathematician is one which no perfectly reasonable man would elect to live.[1]
G H Hardy

This gloomy conclusion was arrived at in 1920 by the most renowned British pure mathematician of the twentieth century – perhaps the only one whose name is known to some non-mathematicians – notwithstanding that during the previous decade he had achieved international recognition as one of the leading practitioners of his generation. Until the Victorian era, pure mathematicians had been an unknown species in Britain, yet by the end of the nineteenth century, when Hardy was a new graduate at Cambridge, they were rapidly becoming an essential constituent of the mathematical scene. Their success did not come easily, because they had adopted a philosophy concerning mathematics that was opposed to the prevailing utilitarianism of the Victorians, and consequently they became the target of considerable criticism concerning the value of their activities. However, this criticism came not only from outside; there was also a good deal of self-criticism and soul-searching, as Hardy, with his usual directness, made so painfully apparent. In this book we investigate the genesis of pure mathematicians in Britain, the cause of their difficulties, the steps that they took to address them, and the eventual alignment of their objectives with one of the most controversial aspects of Victorian popular culture. In the process we shall touch upon some of the most interesting nineteenth-century debates concerning mathematics, which raised questions that are still significant today: What is the purpose of mathematics and why should it be studied and researched? What is the criterion of mathematical truth? What duty to society does a mathematician owe? What value for society does a mathematician have?

Pure mathematics has a long history that goes back to ancient times. It began with the properties of numbers and geometrical figures – Euclid's *Elements*, which dates from about 300 BCE, deals with both – and then expanded to include, amongst other things, algebra, coordinate geometry and the calculus. Such mathematics is described as 'pure' because its concepts do not directly refer to things in the world, or to measurable quantities; rather, these are the province of what is now referred to as 'applied mathematics', which puts pure mathematics to work in a great many scientific, technological and commercial disciplines. To take an example from Newtonian mechanics: to calculate the time taken for a body under constant acceleration to travel a given distance, it is necessary to solve a quadratic equation, and finding a general method of solving quadratic equations is an exercise in pure mathematics because it involves nothing but algebra. Determining that the relation between the four relevant quantities – initial speed, constant acceleration, time taken and distance travelled – can be expressed by a particular quadratic equation, and the use of the general method of solution in specific instances, form part of applied mathematics.[a]

Although Isaac Newton died in 1727, a veneration of his memory continued well into the nineteenth century, and supported a British tradition in which a mathematician's duty was to use pure mathematics for the advancement of scientific and technological disciplines, and to increase our knowledge of the world. Consequently, within the sciences, the value of pure mathematics was believed to lie in its being the essential component of what was then called 'mixed mathematics', a term for which the modern 'applied mathematics' is almost, but not quite, a synonym. This led to a description of pure mathematics as being both 'queen' and 'servant' of the sciences: 'queen', because it was supposed to yield truths greater and more secure than those of the sciences, and 'servant', because its function was to serve the sciences. However, some pure mathematics had no role to play in mixed

---

[a] The equation is $at^2 + 2ut - 2s = 0$, where a is the constant acceleration, t is the elapsed time that it is desired to find, u is the initial speed and s is the distance travelled. When a, u and s are known, the equation can normally be satisfied by two distinct values of t, only one of which will be appropriate for the physical situation.

mathematics, and beginning early in the nineteenth century a few British mathematicians came to believe that it was equally worthy of study, notwithstanding its apparent lack of utility; their commitment to the cause of mixed mathematics showed signs of weakening. Such mathematicians, whatever their personal predilections, at first still had to work in the mixed-mathematical tradition, but later in the century, as circumstances changed, they became able to dedicate themselves to pure mathematics, unconcerned as to whether their work had any relevance elsewhere. This new breed of mathematicians, who wished to devote their time and energy to what many thought of as useless mathematics, became known as 'pure mathematicians', and what distinguished them was not that they studied pure mathematics – which was nothing new – but that their motivation for studying it no longer lay in its current or potential applications.

Until the 1830s, the sciences were brought together as the constituents of natural philosophy, which had the aims of investigating and understanding better the workings of the world, and advancing the fortunes of humankind – all to the glory of God. In this enterprise, mathematics was an important adjunct, and the wide range of its applications meant that use of the word 'mathematician' to describe someone who was not necessarily a specialist, but simply had sufficient knowledge of mathematics to make regular use of it, was common and lasted well into the nineteenth century. In Continental Europe there was an alternative, more abstract approach to mathematics, but communication between British and Continental mathematicians had almost ground to a halt, not least because of differing views that arose in the seventeenth century as to whether the German philosopher Gottfried Leibniz should be regarded as a co-discoverer, along with Newton, of what is now always referred to as 'the calculus'. The British believed not only that Newton should take all the credit, but also that neither the method of fluxions and fluents (as his version of the calculus was called), nor the academic system in which he flourished and produced his great works, could be significantly improved. The wars with France and the excesses of the French Revolution also did nothing to commend Continental ways of thought. Consequently, British mathematicians saw no need to follow developments on the Continent, and in time became almost incapable of understanding them, even had they wished to do so, because of new notations and methods of reasoning that they had ignored. Only at the beginning of the nineteenth century did relations begin to thaw as some mathematicians came to appreciate the significance of Continental advances, which by the 1820s had even begun to find their way into the university curriculum at Cambridge, then regarded as the nation's mathematical powerhouse. During the next two decades traditionalists offered some resistance to this trend, which they saw as concentrating too much on abstract symbolism, and in the 1830s and 1840s there was much disquiet concerning the direction in which mathematics was heading at the university.

Notwithstanding these debates, mathematics was rapidly gaining in importance as an explanatory and descriptive tool. Natural philosophers in all the sciences were acquiring great quantities of data and dealing with theories of ever-increasing complexity and scope, but often found it difficult to reconcile their findings with philosophical and theological beliefs that were based on the teachings of Christianity and the Bible. These circumstances began a process in which the old overarching concept of natural philosophy was eventually abandoned in favour of giving much more independence of thought to each of the sciences – a very welcome consequence for those investigators who wanted to eliminate the more speculative elements from the theories with which they worked.

To further the freeing of science from the shackles of philosophy and theology, in the early 1830s the word 'scientist' was coined and introduced into the language to replace 'natural philosopher', in the hope that it would emphasise the role of rational thinking and hard evidence. It is no coincidence that other changes in vocabulary occurred at the same time: within two decades not only 'scientist' but also the new terms 'pure mathematician', 'applied mathematician' and 'applied mathematics' had all come into regular use, with 'mixed mathematics' beginning to fall out of favour, along with 'natural philosopher'.[b] Although 'scientist' had been preceded by other similar-sounding words such as 'sciencist', 'sciencer', 'scientiate' and 'scientman' – which are now obsolete or rarely used – its invention was the outcome of a discussion between like-minded individuals at a meeting of the British Association for the Advancement of Science, who felt that the language as then constituted did not have a word that identified them collectively as adherents to the new evidence-based regime.[2] Although

---

[b] For the recent introduction of the term 'pure mathematician', see the chart in Figure 5.1 on p. 146.

philosophers, theologians and the clergy were not excluded as scientists, those presenting themselves under this new banner thereby proclaimed their avoidance of the philosophical and theological speculation that had previously permeated scientific enquiry; no longer was the Book of Nature to be read in parallel with the Good Book.

The desire to place the sciences on a completely rational footing gave additional value to mathematical reasoning. This became particularly so after the Great Exhibition of 1851, when a decline began in the relative superiority of the British Empire's products as against those from other countries, due to the prevalence of 'rule of thumb' methods in manufacturing that lacked the precision and efficiency of new techniques that had been developed elsewhere. To remain competitive, manufacturers needed to innovate, and mathematicians were expected to play their part by ensuring that their work supported the nation's attempt to maintain its place in the forefront of new scientific, industrial and commercial developments.

However, that view of the role of the mathematician was not always accepted in Continental Europe, where many believed that mathematics should provide a field for intellectual speculation, rather than serve the sciences. One of the great mathematicians of the age, the German Carl Jacobi, wrote in 1830 to Adrien-Marie Legendre:

> It is true that M. Fourier held the opinion that the main aim of mathematics is public utility and the explanation of natural phenomena; but such a philosopher should have known that the sole purpose of science is the honour of the human spirit, and that under that title a question about numbers is worth as much as a question about the system of the world.[3]

More than a decade later this idea was still a novelty in Britain, as can be judged by Jacobi's experience in 1842 when he was in Manchester to represent Prussia at a meeting of the British Association for the Advancement of Science; later he wrote to his brother Moritz: 'There I had the courage to declare that it is the honour of science to be of no use, which provoked an emphatic shaking of heads.'[4] This reaction should have been no surprise to Jacobi, given that British mathematics and science were principally valued for providing vital support to the country's burgeoning industry, manufacturing, trade and commerce; his audience, situated in Britain's industrial heartland, naturally found his views very unappealing.

As this suggests, the first decades of Victoria's reign were not ones that gave a ready welcome to new, abstract, speculative and inapplicable mathematics. Nevertheless, opinions such as Jacobi's encouraged those British mathematicians who wanted the freedom to study pure mathematics without intending to advance the sciences, or having any reason to think that something useful would come of it. These were the new 'pure mathematicians', a description that soon led to their colleagues who continued working in the old tradition of mixed mathematics being separately categorised as 'applied mathematicians'. Pure mathematicians, who accepted pure mathematics as queen but not as servant, were immediately faced with two difficulties: the first was how best to deflect criticism, of which there was a good deal; the second was how to justify in their own minds their unwillingness to make meaningful contributions to anything that the world would find worthwhile. These difficulties had to be addressed in the face of the prevailing Victorian belief that those with technical skills should deploy them by doing useful work, a belief that was, of course, in direct opposition to the pure mathematicians' ideals.

Consequently pure mathematicians faced an uphill task; but as regards the first difficulty they eventually succeeded, because by the end of the First World War they were well-established in academic life, and had even obtained a kind of moral superiority over their worthy but workaday colleagues, the applied mathematicians. When Hardy wrote in 1940 that 'A mathematician need not now consider himself on the defensive',[5] he was remembering his own experience at Cambridge 40 years earlier. However, the second difficulty – that of self-doubt – remained, and some pure mathematicians found that indulging their predilections did not always sit well with their conscience. At first they were not completely divorced from applied mathematics, and certainly did not in any way seek to diminish its interest and importance; thus they were able to maintain a foot in each camp, and only towards the end of the nineteenth century did a new generation of hard-line pure mathematicians, of whom Hardy was one, became totally committed to pure mathematics. Given this commitment, the sombre comment at the head of this chapter is a surprising reflection on the life that he had chosen for himself, particularly as it formed part of his inaugural lecture as Oxford University's Savilian Professor of Geometry; the explanation is that, although his work gave him much pleasure, he was also very aware of how little he contributed to the common good, and this knowledge induced in him a profound feeling of melancholy.

There is an interesting parallel, evident to the Victorians themselves, between the situation in which pure mathematicians found themselves during the second half of the nineteenth century, and the situation of classicists. They both appeared to become increasingly irrelevant; indeed, pure mathematicians were probably the more irrelevant of the two, for although some proficiency in Latin and ancient Greek (very different from modern Greek) was no longer always expected from those who were well-educated, it was still necessary for the study of ancient history and literature, the value of which was appreciated by all. On the other hand, pure mathematicians apparently provided no benefit to society. Nevertheless, classics continued to decline as an academic subject, and with it the standing of classicists; but the standing of pure mathematicians steadily increased until, in the first decade of the twentieth century, Cambridge produced two of the world's finest, in Hardy and his frequent collaborator John Edensor Littlewood.

This unexpected elevation of the status of pure mathematicians, who finally abolished any hint of servitude from their motivations, brought about what some still consider to be unwelcome consequences that are illustrated by a comment from the French–American algebraist Serge Lang in the 1980s. When asked what algebra was good for, he replied 'It's good to give chills in the spine to a certain number of people, me included. I don't know what else it is good for, and I don't care.'[6] Such a lack of concern for useful applications led Achim Bachem, a distinguished German mathematician, to be 'very worried that mathematicians are moving further and further away from the relevant problems in the natural sciences and society'. His diagnosis was that real-world problems are too messy, and have too many complications, to be truly satisfying for a mathematician; but after adding that 'it is unfortunate that the majority of good mathematicians are interested in pure mathematics', he suggested that 'a mathematician would like to be recognized as a capable professional within his own community. But nowadays, he can only achieve this by asserting himself within the classical value system of ever deeper theorems. If he works with real problems, then he cannot present his result as a deep theorem.'[7]

In the following chapters, we shall investigate how this came about: how pure mathematicians and 'ever deeper theorems' gained an ascendancy at a time when the prevailing culture emphasised utility and pragmatism, and in the face of the well-established British tradition of mixed mathematics. We do so by treating mathematics, not as a body of knowledge, but as a practice – as something that people *did* – and that was therefore subject to public scrutiny and judgement. Accordingly, our account stresses pure mathematicians in society, rather than the content of pure mathematics itself, for they did not work in a vacuum, driven only by the remorseless logic of their theorems; whether they liked it or not, they were social animals, and therefore hoped to work in a world that, if not enthusiastic, was at least acquiescent and uncritical, letting them pursue their vocation without interference.

The actual content of mathematics is not absent from this story, but it does take a back seat. During the nineteenth century, mathematicians made a fundamental re-evaluation of the criteria by which mathematical truth was judged, and Bertrand Russell, who was a mathematician before turning to philosophy, made the point in a typically extravagant fashion when he wrote in 1901 that:

> The nineteenth century, which prided itself upon the invention of steam and evolution, might have derived a more legitimate title to fame from the discovery of pure mathematics. This science, like most others, was baptised long before it was born; and thus we find writers before the nineteenth century alluding to what they called pure mathematics. But if they had been asked what this subject was, they would only have been able to say that it consisted of Arithmetic, Algebra, Geometry, and so on. As to what these studies had in common, and as to what distinguished them from applied mathematics, our ancestors were completely in the dark.[8]

Formerly, and in accord with the tenets of natural philosophy, mathematics was true because it derived from, and aligned itself with, our perception of the world. Latterly, mathematical truth meant logical consistency, which Russell believed to be a unifying concept that created what was in effect a new discipline. However, although the criterion was now that of logic, logic alone was not enough; it needed foundations on which to build. Unfortunately, as Russell was to find out shortly afterwards, the insufficient attention that mathematicians had paid to underlying principles meant that some of those principles led to paradoxical and contradictory results, particularly when dealing with the infinite.

This added to the already well-established abstract and speculative character of pure mathematics, and so diminished still further the ability of pure mathematicians to connect with anyone other than their own kind, for by then it was not possible to make even a spurious claim of relevance to anything that the world-at-large might find of value.

On the other hand, the speculative creations of pure mathematicians may *later* have unexpected and unintended applications. For example, after the publication in 1905 of his paper on the electrodynamics of moving bodies (usually referred to as the special relativity paper), the German physicist Albert Einstein needed a mathematical method capable of handling the ideas that he had developed for his general theory of relativity. He was not a mathematician by training, but his collaborator Marcel Grossmann was, and *he* knew about the tensor calculus, a recent speculation by a group of Italian pure mathematicians.[9] Grossmann realised that it was the ideal vehicle for expressing Einstein's ideas in a mathematical form, and since then it has also found many other applications in physics and engineering. There are other examples: the theory of matrices, which was substantially created by Arthur Cayley as an exercise in pure algebra, is now used in almost every branch of applied mathematics; and the theory of numbers, which was a special interest of Hardy's, and believed to have no conceivable use at all – that was one of its attractions – is now vital in maintaining the security of electronic data and communications; Hardy would have been very disappointed. However, although there could always be a remote possibility that, at some future time, this or that development in pure mathematics might provide benefit to the world-at-large, it was not possible to run the argument that, on those grounds and for no other reason, pure mathematicians should be allowed free rein; what we would now call the cost–benefit ratio was too great. Yet they managed to tap into an aspect of Victorian culture that worked in their favour, produced the desired result, and continues to colour contemporary attitudes towards them and their work.

In the following chapters, we shall see how, over the course of the nineteenth century, pure mathematicians came into being, how difficult it was for them to survive in a hostile environment, and how they achieved their eventual success. Yet it is not only mathematicians and mathematics that are considered, for the narrative is embedded in the world of Victorian Britain, and demonstrates how the adherents of a discipline that appeared to be so removed from the influences of contemporary culture could nevertheless use that culture to forge a role for themselves that achieved their overt ambitions, although not always their mental quietude; in this process, advances in mathematical knowledge – whatever that is taken to mean – played little part. Pure mathematicians had to justify their claim to a place in public life, and obtaining acceptance and respect, as they eventually did, was by no means straightforward. In following their progress, we shall be witnessing the emergence of a non-utilitarian discipline in a highly utilitarian age.

Notes and References

1. G H Hardy, *Some Famous Problems of the Theory of Numbers, and in particular Waring's Problem: An Inaugural Lecture Delivered Before the University of Oxford* (Oxford: Clarendon Press, 1920), p. 4; he gave the lecture on 18 May 1920.
2. [Anon.] William Whewell, 'On the Connexion of the Physical Sciences. By Mrs Somerville', *The Quarterly Review*, LI, CI (1834), 54–68, at 59.
3. 'Il est vrai que M. Fourier avait l'opinion que le but principal des mathématiques était l'utilité publique et l'explication des phénomènes naturels; mais un philosophe comme lui aurait dû savoir que le but unique de la science, c'est l'honneur de l'esprit humaine, et que sous ce titre, une question de nombres vaut autant qu'une question du système du monde.' From a letter dated 2 July 1830: C W Borchard (ed.), *C. G. J. Jacobi's Gessammelte Werke*, 1 (Berlin: G Reimer, 1881), pp. 454–455.
4. 'Ich hatte den muth dort den satz geltand zu machen es sei die ehre der wissenschaft keinen nutzen zu haben, was ein gewaltiges schütteln des kopfes hervorbrachte.' From a letter dated 25 September 1842: W Ahrens (ed.), *Briefwechsel Zwischen C. G. J. Jacobi und M. H. Jacobi: Abhandlungen zur Geschichte der mathematischen Wissenschaften mit Einschluss ihrer Anwendungen*, XXII (Leipzig: Teubner, 1907), p. 90.
5. G H Hardy, *A Mathematician's Apology* (Cambridge: Cambridge University Press, 1967), p. 64; originally published 1940.
6. Serge Lang, *The Beauty of Doing Mathematics: Three Public Dialogues* (New York: Springer, 1985), p. 49.
7. Achim Bachem, 'Mathematics: From the Outside Looking In', in Björn Engquist and Wilfried Schmid (eds.), *Mathematics Unlimited: 2001 and Beyond* (Berlin: Springer, 2001), pp. 275–281.
8. Bertrand Russell, 'Mathematics and the Metaphysicians', in *Mysticism and Logic and Other Essays* (London: Longmans, 1918), p. 74; originally published 1901.
9. Alicia Dickenstein, 'About the Cover: A Hidden Praise of Mathematics', *Bulletin of the American Mathematical Society*, ns 46, 1 (2009), 125–129, at 126; she quotes a translation of Einstein's manuscript in which he records his indebtedness to Grossmann and other mathematicians.

2 THE LEGACY OF NEWTON

Four centuries ago there was a very different understanding of mathematics from that which we have today. Throughout the Middle Ages, mathematics had comprised the ancient *quadrivium* of arithmetic, geometry, astronomy and music, which together with the more elementary *trivium* of grammar, rhetoric and logic constituted the seven liberal arts, as taught in European universities. The broad heads of the *quadrivium* embraced diverse areas of study, and in 1570 the mathematician, astrologer and occultist John Dee gave a comprehensive and influential account of what he considered was the proper scope of the 'Sciences, and Artes Mathematicall'. His scheme was founded upon arithmetic and geometry, from which sprang applications that included astronomy, music, astrology, geodesy, geography, cosmography and navigation – not to mention 'the great Arte of Algiebar'. Because of the value of mathematics in a wide range of practical, non-academic occupations, Dee did not want its study to be confined to the universities, and so wrote in English rather than Latin.[1]

Dee's portrayal of mathematics is confirmed by the reminiscences of John Wallis, who held the Savilian chair of Geometry at Oxford University from 1649. In 1697 he looked back to his youth in the early 1630s and recalled that:

> Mathematicks, (at that time, with us) were scarce looked upon as *Academical* Studies, but rather *Mechanical*; as the business of *Traders, Merchants, Seamen, Carpenters, Surveyors of Lands*, or the like; and perhaps some *Almanack-makers in London*....

> For the study of *Mathematicks* was at that time more cultivated in *London* than in the Universities.[2]

At that time, 'mathematicks' was used as a grammatical plural, and often took the definite article; thus 'the mathematicks' were the sciences and arts that Dee and Wallis described. Isaac Barrow, who in 1664 was appointed the first Lucasian Professor of Mathematics at Cambridge, referred to these as 'mathematical sciences',[3] as did many of his contemporaries.

Justified by the practice in ancient Greece, there were deemed to be two categories of mathematics, which became known as pure (or abstract), and mixed (or concrete).[4] The term 'pure mathematics' was given currency in the language by Francis Bacon, the British philosopher, courtier and ill-fated Lord Chancellor, who in 1605 included in *The Advancement of Learning* a succinct and prescient paragraph in which he explained the categories of mathematics:[a]

> The MATHEMITCKS are either PURE, or MIXT: To the PURE MATHEMATICKS are those Scieces belonging, which handle *Quantitie determinate* meerely severed from any Axiomes of NATURALL PHLOSOPHY: and these are two, GEOMETRY and ARITHMETICKE, The one handing Quantitie continued, and the other dissevered. MIXT hath for subiect some Axiomes or parts of Naturall Philosopie: and considereth Quantitie determined, as it is auxiliarie and incident unto them. For many parts of Nature can neither be invented with sufficient subtiltie, nor demonstrated with sufficient perspicuitie, nor accommodated unto use with sufficient dexteritie, without the aide and interveyning of the Mathematicks: of which sorte are *Perspective, Musicke, Astronomie, Cosmographie, Architecture, Inginarie,* and divers others. In the *Mathematicks*, I can report noe deficience, except it be that men doe not sufficiently understand the excellent use of *the pure Mathematicks*, in that they doe remedie and cure many defects in the Wit, and Faculties Intellectuall. For, if the wit bee to dull, they sharpen it: if to

---

[a] While the original spelling has been retained in this quotation, for the convenience of the reader the following silent amendments have been made to the typography: ſ, which is a long s and nowadays often misread as f, has been printed as s; the use of u for v in the middle of a word has not been followed; and likewise the use of v for u at the beginning of a word. Thus Bacon's 'vſe' is here printed as 'use', and 'ſeuered' as 'severed'.

wandring, they fix it: if to inherent in the sense, they abstract it. So that, as Tennis is a game of noe use in it selfe, but of great use, in respect it maketh a quicke Eye, and a bodie readie to put it selfe into all Postures: So in Mathematickes, that use which is collaterall and intervenient, is no lesse worthy, then that which is principall and intended. And as for the *Mixt Mathematickes* I may onely make this prediction, that there cannot faile to bee more kindes of them, as Nature growes furder disclosed.[5]

The provinces of geometry and arithmetic were magnitude and multitude respectively; mathematics that included additional concepts – because it was concerned with the magnitude or multitude of things in the world – belonged to the other category, mixed mathematics, which was the mathematics that would be of use, or further our understanding of creation.

This separation into two categories refers only to the nature of the mathematics itself, and says nothing about the motives of the mathematician. However, mathematicians were always trying to put mathematics to use, so work in pure mathematics was mostly undertaken with that aim in mind. Consequently, the label 'mathematician' was applied very widely and did not imply a specialist in pure mathematics. Thus, we find Narcissus Luttrell writing in 1687 that 'Sir Samuel Morland, the mathematician, is lately married, and to one not of the best reputation',[6] for Morland had varied interests, being not only a spy, cryptographer and diplomat, but also the inventor of many mechanical devices, including pumps and calculating machines. As this suggests, a mathematician was simply a person engaged to some extent with the mathematical sciences, and therefore skilled, as Morland was, in their practical application.

This view of mathematics and mathematicians continued into the eighteenth century, and so in 1755 Dr Johnson, who treated 'mathematicks' as a collective noun, included in his *Dictionary*:

Mathematicks: That science which contemplates whatever is capable of being numbered or measured; and it is either pure or mixt: pure considers abstracted quantity, without any relation to matter; mixt is interwoven with physical considerations.[7]

In 1761 Thomas Walter, a 'teacher of mathematics', took a similar line in his *New Mathematical Dictionary*, and throughout the eighteenth century the terms 'pure mathematics', 'abstract mathematics' and 'mixed mathematics' were widely used with these meanings. The first two were synonymous, but 'mixed' had no generally received synonym, although 'concrete' and 'compound' are occasionally found.[8]

Mathematical work was categorised as mixed or pure according to whether it involved natural quantities, and this was an objective assessment that could be determined from the work itself. However, there was a further distinction that could be made between the speculative and the practical, a distinction that Isaac Barrow explained and then condemned:

> For, in my Opinion, every Science is both Speculative and Practical: Speculative, as it speculates, i.e. seeks, investigates and demonstrates Truths (or true Propositions) agreeable to its Object: and Practical, as those Truths when found and demonstrated, may be referred to Use, and reduced into Practice. ... Geometry is one simple Science which may be considered in two Respects, viz. it is stiled Speculative as it is true, and Practical as it is useful. The Absurdity then of this Division is clear.[9]

He argued that the sciences and mathematics had as part of their very essence a practical purpose, and did not exist simply to proclaim abstract truths, so the proposed distinction was false and pointless. This was also the general opinion, and throughout the eighteenth century it is rare to come across any reference to 'speculative' or 'practical' mathematics, although Thomas Walter included them briefly in his *Dictionary*.[10]

Much more recently this has been highlighted by Ruth and Peter Wallis, who compiled a list of more than 19,000 eighteenth-century British mathematicians and showed that almost all of them were engaged in a practical trade or vocation; in particular, specialist teachers of mathematics were rare, because most mathematicians learned their mathematics 'on the job' when training for their principal occupation.[11] Although some of the mathematics might be 'pure', there was always an underlying purpose, as explained in 1755 by Peter Shaw, who wrote that it would:

> ... be absurd to spend our time in the bare study of logic, or mathematics; which are only means of our rising higher,

and preparing us for the nobler sciences. Hence it cannot be necessary, for the promoting of human happiness, that the world should abound with mere logicians, or mere mathematicians...[12]

Ruth and Peter Wallis have shown that Britain, at least, certainly did not abound with 'mere mathematicians'. However, the absence of pure mathematicians does not mean that mathematics made no progress as a subject for academic study: the situation described by John Wallis did not last.

From the 1750s the posthumous influence and reputation of Newton ensured that at his university of Cambridge mathematics began to oust classics as the discipline in which high-flying students would try to excel: until 1824 the only Tripos, or BA honours examination, was in mathematics.[b] The practice of publicly ranking successful candidates in order of merit generated intense competition, and the senior wrangler became a national figure, to the extent that Augustus De Morgan quoted a contemporary maxim: 'Oxford shall settle what the world shall think, and Cambridge shall settle who is to be Senior Wrangler'.[13] Thus, a clear divide opened up between the two universities of Cambridge and Oxford, for at the latter, which had no one to rival Newton amongst its alumni, classics and theology continued to be the principal areas of study, and there was far less overt competition between students. Classics also remained the main preoccupation of schools, and even some of the best-known did not consider mathematics to be an academic subject, rarely teaching it beyond basic numeracy. This is clear from the evidence collected by the Clarendon Commission, which had been set up by the government to enquire into (amongst other things) the curriculum and teaching in certain colleges and schools. It issued its report in

---

[b] Cambridge had its own terminology for successful candidates in the (Mathematical) Tripos. Those obtaining the equivalent of first-class honours were termed 'wranglers', from the former practice of holding public disputations, or wrangles. The top-placed candidate was the senior wrangler; others were referred to by their ranking, e.g. fifteenth wrangler. Candidates obtaining the equivalent of second- or third-class honours were referred to as senior optimes and junior optimes, respectively; thus a candidate might be referred to as the fifteenth senior optime. The lowest-ranked junior optime was 'the wooden spoon', and students who were studying only for a pass degree were οἱ πολλοί, or poll men. Only men could be declared wranglers; the first woman allowed to take the Mathematical Tripos was Charlotte Angas Scott in 1880, but her ranking had to be defined as being 'between the seventh and eighth Wrangler'. In 1890 Philippa Fawcett became the first woman to be placed 'above the Senior Wrangler'.

1864 and noted that 'the study of mathematics was first made compulsory at Harrow in 1837'; it was also very critical of Eton, where 'before the year 1836 there appears to have been no mathematical teaching of any kind', although elementary arithmetic was taught.[14]

It is therefore unsurprising that the mathematics studied at Cambridge remained very much in the Newtonian tradition, but that alone is not the reason for a long hiatus in communication between British and Continental mathematicians that lasted for the whole of the eighteenth century. There is no single explanation for this, but an important factor was a dispute that started late in the seventeenth century concerning the calculus and its origins – the word 'calculus' is from Latin and refers to stones or pebbles that in ancient times were used to help with arithmetic. Nowadays 'a calculus' can be a scheme or method of calculation, but without any other indication 'the calculus' is shorthand for mathematical methods that, at their simplest, deal with instantaneous rates of change (the differential calculus) and areas bounded by curves (the integral calculus). In the early days of the calculus it involved nebulous concepts such as 'infinitely many', 'infinitely small' and 'quantities that did not exist but were being generated', over which there was much understandable confusion; and although Newton and Leibniz each produced a workable system that yielded valuable results, the two systems appeared to be very different, so that even today historians disagree as to the extent to which they are, or are not, fundamentally the same. Whereas Newton's was called the theory of fluxions and fluents, Leibniz's terminology was similar to that now used, and our modern understanding of the calculus is founded on his system and symbolism, suitably updated. However, for nearly a century after Newton's death there was intense nationalistic feeling on both sides of the Channel, not only as to the relative merits of the two systems, but also as to whether Leibniz, when formulating his system, had already been aware of Newton's work. Only slowly, and with reluctance, did the British come to realise that, regardless of the systems' origins, Leibniz's was superior, and should be adopted.

This difference of opinion, in which the British thought that they were defending the genius of Newton, lasted for the whole of the eighteenth century, and gave rise to their belief that they had nothing to learn from the Continent; so with no perceived need for communication with Continental mathematicians, Cambridge University soon became a mathematical world unto itself. Consequently British mathematicians had great difficulty in understanding the work of their Continental counterparts, even when they wished to do so; for example, by the 1770s Edward Waring, the Lucasian Professor at Cambridge, found it virtually impossible to follow the work in mechanics of Leonhard Euler, the Swiss who was perhaps the greatest mathematician of his generation.[15] Likewise, the Continentals were rarely concerned with what was going on in Britain, and the French mathematician Jean de la Ronde d'Alembert said in 1769 'that if an Englishman is to be elected one of the eight foreign associates of the Academy of Sciences, he will vote for the Earl of Stanhope as the best mathematician there, as he believes, not having read any of his works!'[16] However, it was on the Continent that for a century the most far-reaching mathematical advances would be made, so Continental mathematicians lost far less than the British through this breakdown in communication.

In the opening years of the nineteenth century a few British mathematicians came to believe that it was essential for their country's mathematical isolation to end, and so they made it their business to study the methods developed on the Continent during the previous century. At Cambridge, James Toplis produced a partial annotated translation of Laplace's famous work on mathematical astronomy *Traité de Méchanique Céleste*, and Robert Woodhouse wrote a textbook that introduced the notation for the calculus developed by Leibniz. However, the time was not yet ripe for these novelties, and the books made little impact – so little that it was left to Mary Somerville, nearly 30 years later, to fill the astronomical gap with *Mechanism of the Heavens*, her widely praised rewriting, in English, of Laplace's original text. As for the calculus, Sir Frederick Pollock, the senior wrangler and first Smith's prizeman of 1806,[c] wrote to Augustus De Morgan in 1869 saying: 'I consider that I was the last *geometrical* and *fluxional* senior wrangler; I was not up to the *differential* calculus, and never acquired it'; the references were to Newton's and Leibniz's systems. De Morgan replied with the names of five later senior wranglers who 'may contest this point with you'; the most recent was Cornelius Neale, the senior wrangler and first Smith's prizeman of 1812.[17] Given that these were the

---

[c] In 1768 the will of Robert Smith instituted two Smith's Prizes for mathematics, which were competed for annually – usually by the most successful recent wranglers. The style of the questions differed from that of the Tripos by giving candidates greater scope for originality and creativity. The prizes were equal in value, but the winning candidates were ranked first and second.

most accomplished of the Tripos candidates, it is clear that at Cambridge ignorance of Leibniz's calculus lasted well into the second decade of the nineteenth century.

After the efforts of Toplis and Woodhouse it was left to a handful of Cambridge undergraduates to make a more determined attempt at introducing Continental mathematics by forming the Analytical Society in 1812; it existed for not much more than a year, with a maximum active membership of about 16.[18] In 1813, it published a volume of memoirs, and in 1816 three of the most active former members – John Herschel, George Peacock and Charles Babbage – produced a translation of *Sur le Calcul Différentiel et Intégral* by the French mathematician Silvestre François Lacroix. The extent of the society's influence is unclear and has been much debated, but there is no doubt that thenceforward the Leibnizian notation for the calculus became increasingly familiar, allowing tutors and students alike to understand Continental texts – if they could grapple with the language. By 1820, this trend had led George Biddell Airy, a future Astronomer Royal then in his first year at Cambridge, to learn French 'as I perceived that it was absolutely necessary for enabling me to read modern mathematics'.[19] During the 1820s, textbooks and examinations increasingly reflected the new Continental methods,[20] and in 1831 John Wright, a Cambridge-trained pedagogue, referred to the French analytical system 'now so firmly established' at Cambridge.[21]

At its most obvious, the new system concerned notation. Amongst other things, the calculus gives us rules for calculating rates of change, which Newton referred to as fluxions, and in modern terminology are known as derivatives or differential coefficients. The notation developed by Newton placed dots over expressions to indicate the fluxion. To take a simple example, if the distance of a moving object with respect to some origin is given by a formula x that includes a reference to the elapsed time t, then the object's speed – that is to say, the instantaneous rate at which distance is changing with time – would be written as $\dot{x}$, and its acceleration – which is the instantaneous rate at which speed is changing with time – would be written as $\ddot{x}$. British mathematicians had grown used to this method of notation, and it is still used occasionally in such straightforward cases. However, when complicated expressions were involved, using dots could be very awkward. According to early enthusiasts for the Continental system, dots were also ambiguous and unsuggestive;

consequently they advocated the Leibnizian notation, which introduced an explicit reference to the elapsed time t, and in which speed and acceleration would be written as $\frac{dx}{dt}$ and $\frac{d^2x}{dt^2}$ respectively. This looks more involved than the dot notation, but in use has many advantages, as is clear from the speed with which most British mathematicians finally adopted it, once they had overcome their instinctive aversion to a Continental practice.

At a deeper level, however, the debate introduced metaphysical issues that were to permeate British mathematics for the rest of the nineteenth century, and beyond. In Newton's theory of fluxions the underlying concepts are geometrical, as Thomas Walter tried to explain in his *Dictionary* – not with complete success:

> Fluxions, in mathematics, are the *velocities* of the increments of variable or indeterminate quantities, considered not as actually generated, but as arising or beginning to be generated: thus a line is described by the constant motion of a point. A *surface* by the uninterrupted motion of a line, and a solid by the constant motion of a superficies.[22]

It is difficult to imagine an increment of a variable or quantity 'not as actually generated, but as arising or beginning to be generated'; however, the concepts underlying the theory were still being worked out. Nevertheless, it was clear that the doctrine of fluxions linked motion and geometry, and that this allowed the reformulation of dynamical problems in geometrical terms. Although the difficult questions concerning the existence and nature of these quantities were never satisfactorily answered, the whole system was clearly empirical, and constant attention to physical principles should have prevented egregious errors.

This approach sat well with the firmly held British belief that certainty could result only from a style of argument in which the meanings and mutual relations of the terms were kept constantly before the mind, and which the reasoning of the Newtonian tradition was thought to embody to perfection. However, the British also believed that Continental mathematicians, in the approach frequently referred to as 'analytics', increasingly lost sight of the natural meaning of symbols and sought to manipulate them by rote, with the result that developing an argument meant simply applying rules of logic so that one line generated the next. This was anathema to the British, who found it far less rigorous as a demonstration and considered it to be a symptom of mental laziness.

This was a manifestation of a national characteristic that had been noticed two centuries earlier by Blaise Pascal, and early in the twentieth century the French physicist and philosopher Pierre Duhem invoked as still relevant Pascal's distinction between the French and English minds. The French is narrow and deep; it can comprehend only a limited range of phenomena at one time, but then seeks to organise it into a strictly logical abstract deductive system derived from a few clear axioms. On the other hand, the English mind, ample but shallow, can hold many more phenomena in view at the same time without confusion, but is averse to performing the abstraction and logical arrangement that comes so easily to the French.[23]

Nor was acceptance of Continental methods assisted by there being in Europe disagreement as to the definition of the differential coefficient of a function. At its simplest, a function can be understood as a mathematical expression the value of which depends on some other value (the variable). In the previous example, the expression x is a function of the variable t; and the differential coefficient ($\dot{x}$ or $\frac{dx}{dt}$) is another function of t that gives the instantaneous rate at which x is changing. However, while there were agreed rules for calculating the differential coefficient, there were two candidates for the underlying definition, both of which involved infinite series.[d] In one case it was taken to be the limit of a series, whereas the other involved particular terms of a series known as the Taylor expansion. Lacroix had adopted the Taylor definition in his *Traité du calcul différentiel et du calcul intégral* but used the limit in the more elementary work that Herschel, Peacock and Babbage translated. However, in the interests of supposed simplicity they replaced the limit definition by the Taylor definition, so that they could establish the calculus 'upon principles which are entirely independent of infinitesimals or limits',[24] or so they thought. Unfortunately, few

---

[d] A *finite series* consists of a finite number of terms in a particular order, added together; but sometimes it is possible to define a series with no final term, as with $1/2 + 1/4 + 1/8 + \cdots + 1/2^n + \cdots$. Such series are referred to as *infinite series*, and because they cannot be summed in the usual sense the sum as written appears to be meaningless. However, for some infinite series there is a unique number (the *limit*) to which the sum of a finite number of terms (beginning with the first) tends as more and more terms are summed: in the example just given the limit is 1. If an infinite series has a limit, then it is possible to make 'the sum of the series' meaningful by equating it to the limit. In modern algebra, these concepts are expressed symbolically using much more rigorous definitions.

mathematicians shared the necessary assumption that the Taylor expansion could be established without appeal to limits, and 20 years later even Lacroix's compatriots no longer defended it.[25] Certainly the limit definition, rather than the Taylor definition, was more acceptable to the British, and Augustus De Morgan, the professor of mathematics at London's University College, wrote in 1855 that 'Nothing is to me more certain than the facility with which any fluxional proof which is good of its kind can be rendered into the language of limits.'[26] However, in the first half of the century this disagreement as to the definition of the differential coefficient gave mathematicians who instinctively preferred the theory of fluxions a reason to continue with their traditional beliefs, while using the Leibnizian notation merely as a matter of convenience.

Furthermore, the philosophy behind Continental mathematics struck at the heart of the Cambridge concept of a liberal education, in which what mattered was not so much the mathematical knowledge acquired as the process of acquisition. There was almost no career path outside the university that required the kind of mathematics that was taught there; nevertheless, the study of mathematics was believed to inculcate unique powers of reasoning that would be of inestimable value to the student, whatever his final station in life. This belief ensured that after a Classical Tripos was first examined in 1824, for nearly 30 years the only candidates allowed to take it were those who had already been successful in the Mathematical Tripos, or were the sons of peers.[27] Consequently, until the middle of the nineteenth century almost all Cambridge honours graduates, even classicists with no interest in mathematics, had been obliged to acquire a thorough mathematical training of a very characteristic kind. Other institutions that might have been expected to provide a strong counterbalance to the Cambridge tradition did not in fact do so. The output of mathematicians from Oxford and the four Scottish universities was low, and more recent providers of higher education that had appeared during the 1820s and 1830s (such as London University, the University of Durham, King's College (London) and University College (London)) either did not teach mathematics at all, or did so at a level that was not comparable with that of Cambridge. Consequently there was a large pool of eminent Cambridge graduates in all walks of life who were keen to bear witness to the value of their mathematical training, but there were very few others, whether from Cambridge or elsewhere, who were able to offer an alternative point of view.

Given their failure to keep up with the latest developments, it is unsurprising that British mathematicians produced little new work of lasting significance; which partly explains why, although there was an abundance of journals offering mathematical puzzles and diversions, until 1837 there was no specialist mathematical journal in England, and most mathematical papers appeared in the *Philosophical Transactions* of the Royal Society. Other scientific journals of the time were the *Philosophical Magazine*, the *Annals of Philosophy*, and the *Journal of Science and the Arts*, but they published very few specialist mathematical articles;[28] and the two Continental journals of Crelle and Liouville, later to be outlets for British mathematicians, were not founded until 1826 and 1836 respectively. The 760 papers published in the *Philosophical Transactions* between 1800 and 1830 have been reviewed by J M Dubbey, who found that only 35 had original mathematical content by British authors. Dubbey comments that the geometrical papers show that 'British mathematicians were at this time as unfamiliar with Descartes as they were with Leibniz.'[29] The figures are remarkable, for 15 of the 35 papers were contributed by Woodhouse, James Ivory, Babbage and Herschel, who were all enthusiasts for Continental analytics before participating in the Analytical Society; and Thomas Knight, the other notable contributor, was a botanist. This strongly suggests that in the 18 years following the demise of the society its effect on the specialist output of British mathematicians was minimal, and that the notion of undertaking and publishing research into pure mathematics for its own sake was largely unintelligible.[30]

Consideration of the role of the mathematical professors at Cambridge reinforces this conclusion. There were three chairs, and as natural philosophers were deemed to be generalists the predilections of an appointee had little to do with the ostensible subject matter of the chair; only the Lucasian, which had been Newton's, specifically referred to mathematics, and it was usually held by a natural philosopher working in the Newtonian tradition. He was not required to lecture, or even to reside in Cambridge, and Charles Babbage did neither during his 11-year tenure from 1828 to 1839.[31] Richard Sheepshanks, who enjoyed stirring up controversy, later wrote:

> Mr. Babbage never lectured at all, though he once proposed to lecture, and I believe I helped to stop him. He gravely proposed to lecture *immediately after the Senate House examination* [the

Tripos], when there is no one in the University; and the bill of fare was to be composed of what he had written (about the Economy of Manufactures, I believe,) for the *Encyclopaedia Metropolitana*. ... The Lucasian Professorship is very poorly paid, – not better than the Secretaryship of the Royal Society, – and so hampered by statute, that we can only get a good professor by chance; but I do not see what excuse this is for Mr. Babbage.[32]

Robert Woodhouse, the early advocate of Continental analytics, was Lucasian Professor from 1820 to 1822, when he moved to the Plumian chair; although the Lucasian allowed its holder to bask in the reflected glory of Newton, it was also the worst remunerated. The interests of Thomas Turton, Woodhouse's successor as Lucasian, clearly lay elsewhere, for in 1827 he was appointed Regius Professor of Divinity, and subsequently dean of both Peterborough and Westminster, and bishop of Ely.

The Plumian chair was for astronomy and experimental philosophy, and the Lowndean Professor had to lecture on astronomy and geometry, as well as making astronomical observations. The holder of the Lowndean for 42 years until 1837 was the now wholly forgotten William Lax, and for most of that time he was also vicar of two parishes to which he had been presented by Trinity College; at one of the vicarages he built a small observatory. He compiled some tables for the Board of Longitude and published two astronomical papers; his only contribution to pure mathematics was the 19-page *Remarks on a supposed Error in the Elements of Euclid*, issued in 1807. He was succeeded by George Peacock, an early member of the Analytical Society, who held the chair for 22 years, during most of which time he was also Dean of Ely and so responsible for the restoration of the cathedral. Professors were not obliged to retire, and normally held their chairs for life.

On the Continent, life for the holders of professorial chairs was very different. In particular, in France, and to a lesser extent in Prussia, professors were heavily regulated as to the quantity and subject matter of their teaching; they were in effect civil servants operating under the control of ministers, and had none of the freedom of action enjoyed by the holders of chairs at Oxford and Cambridge.[33] However, this did not mean that they were entitled to neglect research, which in Prussia was the basis on which they were evaluated, notwithstanding that their salaries were paid for teaching.[34]

Early in the nineteenth century, concerns that the broadly based British view of the sciences was under threat from the trend towards specialisation gave rise to a vigorous debate within the Royal Society, which had a membership that traditionally comprised gentlemen interested in natural philosophy. Although some members were of scientific eminence, most were not, and until the middle of the century there was no formal entrance requirement, apart from nomination by three Fellows. For many of its members the eclectic nature of the membership was a great strength, and they feared the narrowness of view that specialists would bring. However, some members believed that the rule of England's foremost scientific society by amateurs and dilettantes had had a stultifying effect on British science, and that reform was urgently required. Amongst the most vociferous of the reformers were Herschel and Babbage, and in the 1830 presidential election Herschel was nominated as a reform candidate to oppose the Duke of Sussex, a son of the deceased King George III with no scientific credentials. The Duke won, although narrowly, and shortly afterwards Babbage published a broadside against the manner in which science was prosecuted in England; in particular he lambasted the lack of official support for the abstract sciences, by which he meant primarily pure mathematics. He argued that it was always possible for an apparently purposeless advance in pure mathematics to have some unsuspected application later, and therefore support of the discipline was necessary because of its potential; yet only the possessors of a private fortune could afford to pursue it.[35]

That there had been no significant change in the fundamental ideas held by most British mathematicians concerning the nature of mathematics can be seen from the use of mathematical terms. 'Mathematics', 'pure mathematics' and 'mixed mathematics' were still often construed as plurals, suggesting a collection of techniques and methods rather than a single discipline, and the term 'applied mathematics' was rarely found; its first recorded appearance in English, in 1798, was in a translation of Kant.[36] However, early in the nineteenth century there is evidence that some mathematicians were seeking a narrower definition of mathematics; for example, in 1814 Peter Barlow published a new mathematical dictionary, in which he wrote that:

> ... a number of terms usually introduced into works of this description, have been omitted. Such are all those relating to the exploded science of Astrology, and those which are merely technical in Architecture, Fortification, Music, and Military Affairs; which, though proper articles for an Universal Dictionary, do not necessarily, nor properly, form part of a work professedly mathematical.[37]

Barlow was a self-taught professor at the Royal Military Academy, Woolwich. His view as to the proper scope of a mathematical dictionary may be contrasted with that of Thomas Walter 50 years earlier, in whose dictionary there were many references of the kind that Barlow excluded. Barlow implied that mathematics, as such, should no longer encompass such technical applications, for they belonged in other disciplines.

On the other hand, the next year Charles Hutton, Barlow's colleague at the Academy, emphasised in the second edition of yet another mathematical dictionary the traditional role of algebra, saying that it consists of two parts, namely 'calculating magnitudes or quantities, as represented by letters or other characters', and 'applying these calculations in the solution of problems'. This definition was almost identical to that in the first edition 19 years earlier and reflected the Cambridge approach to mathematical pedagogy.[38] George Peacock, one of the translators of the 1816 edition of Lacroix, was clearly sympathetic to the Continental approach, but nevertheless felt obliged to engage with this peculiarly British aversion to abstract reasoning when he published a collection of practical examples in 1820:

> The principal object of this undertaking is to remedy a defect, which exists in greater or less degree, in nearly all the foreign works on this part of analysis, in which the theory is most frequently exhibited in its most general form, without the illustration which particular examples are calculated to afford: the student is consequently bewildered in the generality of the relations which are presented to his mind, and he either rests contented with a vague and imperfect comprehension of the subject, or abandons the study which becomes repulsive and fatiguing from its obscurity and difficulty.[39]

That the primary function of pure mathematics was still to aid natural philosophers in their investigations was also implied by John Playfair in

1824, when he wrote in the *Encyclopaedia Britannica* that 'the history of pure Mathematics will be first considered, as that science has been one of the two principal instruments applied by the moderns to the advancement of natural knowledge. The other instrument is Experience ... '.[40] Thus Barlow, Hutton, Peacock and Playfair all believed that the value of pure mathematics lay in its being a tool with which problems in natural philosophy and practical science might be more readily tackled and solved. Of these four, only Peacock was at Cambridge, but most of those who studied and taught there shared his opinion; however, they did not feel impelled to make their case publicly, instead believing that discussion of such matters should be kept within the confines of the university.

The first three decades of the nineteenth century therefore saw the fundamental outlook of British mathematicians remain much the same. Nevertheless, most were willing to concede that the dispute over the calculus had been harmful to British interests and should be buried, with honours even. Accompanying this conciliatory frame of mind was a grudging acceptance of the utility of the Leibnizian notation for the calculus, and of the need to read, although not necessarily embrace, the work of Continental mathematicians.[41] Furthermore, the scope of the word 'mathematician' was beginning to signify someone who specialised in mathematics rather than, as in the previous century, someone who simply used mathematics as a problem-solving tool in the sciences, arts and other practical occupations. The effect of this restriction was eventually to exclude as mathematicians many who worked in industry, commerce, the armed forces and other professions and trades.

By 1830 there were some tangible signs of change, for in 1828 London University had opened for business. It was a non-Anglican private educational enterprise that was unable to award degrees, despite its name, because Anglican opposition had denied it a royal charter.[42] However, when the degree-awarding University of London was formally incorporated by royal charter in 1836, London University, after being renamed University College, became one of its first two colleges, the other being King's. The nature of London University, University College and the University of London reflected public disquiet with the two ancient universities of Oxford and Cambridge, which were constitutionally Anglican and demanded conformity, whereas these new institutions were non-denominational and non-residential, and exercised no formal control over the religious or political opinions of students and staff. Although King's *was* an Anglican college, London in 1836 had a university that owed little to Oxford or Cambridge; rather, it was a response both to Anglicans who were dissatisfied with the moral and political lead given by their Church, and to the increasing population of radicals and non-conformists, for whom education and self-improvement were of particular concern.

At the same time as Oxford and Cambridge were losing their status as the only English universities, an indication that British mathematics was becoming more open to Continental influences came from the interest in symbols that was being taken by some reformers at Cambridge and by Augustus De Morgan at the University of London. Now that it was possible for mathematicians to follow Continental texts, a growing enthusiasm for the manipulation of symbols led to fears amongst traditionalists that the aims of a Cambridge mathematical education were being ignored; on this view, mathematics was in danger of being reduced to a purposeless exercise in logic. These concerns were given direct expression in an introductory lecture for mathematical students at Belfast College by John Radford Young, the newly appointed professor of mathematics, in 1833:

> It will be, then, my anxious endeavour to explain clearly, and to illustrate copiously, the various formulas with which mathematical analysis abounds, so that the student may be in no danger, while contemplating the symbols, of losing sight of the things signified, a danger to which the solitary student of modern mathematics is unfortunately too much exposed. It was the characteristic excellence of our older mathematical books, that they abounded in practical illustration of the theories taught. Our modern authors are departing more and more from this admirable system; and it is no unusual thing now, in looking through a mathematical book, even upon a practical subject, to meet with nothing but a bewildering array of symbols, without a single illustration of their meaning, or a single example of their practical application. This is more especially the case with works on the higher Calculus, and on Analytical Mechanics, which are little else than abstract algebraical exercises, furnishing no real information to the reader, nor even any evidence that the writers themselves understand their own formulas.[43]

The continuation of this unwelcome trend was confirmed in evidence given by the Rev. Thomas Gaskin, a tutor at Cambridge, to the Royal Commission on Cambridge University that reported in 1852:

> From 1832 to 1847 the tendency was to become more and more symbolical and analytical. Within that period, pure geometrical reasoning was altogether abandoned, and the attention of the Students, by the questions proposed, was unduly fixed upon the dexterous use of symbols, to the neglect of natural relations in the application of Mathematics to physical subjects.[44]

This came about because Continental mathematicians were widening the scope of algebra in new and troubling ways that extended far beyond the numeral and literal algebra already familiar to British mathematicians, some of whom were adopting these novelties with all the enthusiasm of a convert.

The traditional old, numeral or vulgar algebra, an art the origins of which, Dr Johnson had noted, were 'very obscure',[45] was described in the edition of Bailey's *An Universal Etymological English Dictionary* published in 1763:

> [it] was that of the Ancients, and served only for the Resolution of Arithmetical Questions, and is when the Quantity sought is represented by some Letter or Character, but all the given Quantities are express'd by Numbers.[46]

Of more recent vintage was the new, literal or specious algebra; here is Bailey again:

> [it] is that Method by which, as well the given or known Quantities, as those that are unknown, are severally express'd or represented by Alphabetical Letters; and is generally used for all Mathematical Problems, both Arithmetical and Geometrical.[47]

For example, before the advent of literal algebra a quadratic equation had to be expressed with all the coefficients specified as numbers, as in $3x^2 + 7x - 4 = 0$, and could be solved by the method of 'completing the square'; but this required some mathematical sophistication. However, in literal algebra letters also represent the coefficients. Thus the general quadratic equation could be expressed as $ax^2 + bx + c = 0$, and 'completing the square' yielded a formula that was generally applicable and straightforward to evaluate: $x = \frac{-b \pm \sqrt{b^2-4ac}}{2a}$. Other problems in algebra could be generalised and dealt with by similar means. In consequence, algebraical methods became available to a wider circle of practitioners, because a formula such as that for the quadratic equation could be applied mechanically, using sets of tables if required; the experience needed for understanding and applying the method of 'completing the square' was rendered unnecessary.

In mixed mathematics many physical situations therefore became expressible in general algebraical terms that could be invoked in any particular instance. This was a welcome improvement, for which eighteenth-century mathematicians were duly grateful: Thomas Walter wrote with feeling that literal algebra was 'a great relief to the memory, when obliged to keep several matters, necessary for the discovery of the truth in hand, present to the mind',[48] and Barlow noted that 'all the conclusions become universal theorems for performing every operation of a similar nature with that for which the investigation was instituted'.[49] British mathematicians, taking their lead from Newton, had become perfectly familiar with literal algebra during the eighteenth century, and it caused no difficulty.

However, as Young and Gaskin complained, the tendency of Continental analytics was to concentrate on the manipulation of symbols without regard to the nature of the underlying quantities. This was particularly problematic when dealing with infinite series that involved a variable because, under the definition with which most British mathematicians worked, the sum of such a series existed only for certain values of the variable. However, an enthusiast for analytics did not consider that the value of the variable was a legitimate concern of algebra; either an expression had been correctly obtained from some other expression by the manipulation of symbols, or it had not. If it had, then it was universally valid. The result was an argument such as this: consider the infinite series $1 - x + x^2 - x^3 + \cdots$ (in which the even powers of x take the positive sign, and the odd powers the negative), and the effect of multiplying it by $1 + x$. If algebra is regarded simply as an exercise in manipulating symbols, then it follows that the equation $(1 + x)(1 - x + x^2 - x^3 + \cdots) = 1$ is always true, because on multiplying out the left-hand side we get $1 + x - x + x^2 - x^2 + x^3 - x^3 + \cdots$; clearly, all the terms in x cancel. But if the equation is always true, then so is the result of dividing it by $1 + x$, which yields $1 - x + x^2 - x^3 + \cdots = \frac{1}{1+x}$.

However, the usual understanding of the sum of an infinite series was as a limit to which the series converges; this means that the sum of the infinite series $1 - x + x^2 - x^3 + \cdots$ has meaning only if $-1 < x < 1$, because with other values of x the sum of the series increases without limit or oscillates in a non-convergent manner as more and more terms are included. Those algebraists who supported the first argument, however, were not deterred by this, and championed curious results: for if $1 - x + x^2 - x^3 + \cdots = \frac{1}{1+x}$ is always true, then using, say, $x = 2$ it follows that $1 - 2 + 4 - 8 + \cdots = 1/3$. This, although offensive to common-sense, cannot be disproved by appeal to physical considerations, as could an equation in finite arithmetic, for it is a logical consequence of definitions and assumptions concerning infinite series; but it does show how analytics could lose touch with the belief that the role of symbols, and the principles for manipulating them, needed to be consistent with natural processes. This example is taken from the wonderful polemic against such practices written by Archibald Sandeman, the joint third wrangler and second Smith's prizeman of 1846, who was the first professor of mathematics at Owens College, Manchester. His target was 'algebraers' who raise as a principle 'that a chain of reasoning to be strong and good need not have meaning in every link ... '; he concluded, with a lack of punctuation that was entirely typical of his style, 'Small need then to say as a wind up that arithmetic and algebra in their wonted setting forth cannot but be educationally bad and mischievous scientifically misleading bewildering unhelping balking stunning deadening and killing and philosophically worthless.'[50]

The kind of counter-intuitive arguments offered by some mathematicians immediately raised the question: in what sense, if any, could the results of algebraic manipulation represent truth, if natural relations were no longer reference points? For example, pure algebra suggests that the equation $x + 1 = 0$ has the solution $x = -1$. However, -1 cannot represent any natural quantity, and therefore a few mathematicians did not accept it as a number; for them, the equation $x + 1 = 0$ had no solution. William Frend, Augustus De Morgan's father-in-law, was one such, for he indulged in 'mathematical heresy, the rejection of the use of negative quantities in algebraical operations'.[51] Likewise, in 1823 a contributor to *Encyclopaedia Britannica* wrote that:

> When quantities are considered abstractedly, the terms *positive* and *negative* can only mean that such quantities are to be added

or subtracted; for as it is impossible to conceive a number less than 0, it follows, that a negative quantity by itself is unintelligible.[52]

Despite misgivings, however, the Continental analytic trend had become clear by the early 1830s, and particularly so to William Whewell, later to be a figure of great power and influence at Cambridge: the second wrangler and second Smith's prizeman of 1816, he became professor of mineralogy and Knightbridge Professor of Moral Philosophy, finishing his career as Master of Trinity. Unlike Babbage and De Morgan, Whewell came from a working-class background, possessed great physical energy, and was an Anglican priest. As a young man he supported Continental analytics, referring to it as 'the true faith', but then became increasingly alarmed at the new devotion to symbolic manipulation.[53] Nevertheless, he did not agree that the grounds for believing mathematics to be true were wholly empirical, for intuition played an important role; and so in furtherance of his philosophy, and to try to turn the tide of events at Cambridge, he published in 1835 his *Thoughts on the Study of Mathematics as part of a Liberal Education*, which ran to a second edition the next year.

In this little book Whewell described 'a peculiar school of mathematical speculation',[54] with characteristics that were opposed to the traditional Cambridge understanding of mathematics. He established without much ceremony 'that it is a proper object of education to develope and cultivate the reasoning faculty',[55] and his whole book is premised on this; it may also have been a proper object to impart knowledge, but that was of less importance for the kind of liberal education with which he was concerned. He then asked: how should the universities proceed? In answering, he contrasted the pedagogical merits of mathematics and logic, which were respectively 'the teaching of reasoning by practice and by rule'. In mathematics, the student learned by working through examples of 'strict inference' and considered all the variety of surrounding natural circumstances by which the 'cogency of the demonstration' would be judged. In logic, on the other hand, there were clear formal conditions that defined what was meant by correct inference, and the student's task was the almost mechanical one of confirming that the demonstration complied with the conditions.[56]

Whewell was in no doubt that mathematics was more effective than logic, and he identified his opponents as holding two dangerous opinions. First, they denied the role of 'the faculty of the mind and the intellectual process',[57] whereas he believed that intuition provided certain fundamental ideas such as the straight line and the limit.[58] He also held the traditional view that a sound demonstration required an argument to be comprehended in its entirety by the mind, for which reason he lamented the disappearance of oral examinations, and as late as 1852 considered that examinations conducted wholly on paper were 'a very imperfect method of ascertaining whether the candidates fully understand the subjects'.[59] Second, mathematical speculators hold forth 'as the highest perfection of our knowledge, to reduce it to extremely general propositions and processes, in which all particular cases are included',[60] and on this ground he also condemned some recent university treatises that 'can do no service to the cause of education, if their merit be their generality'.[61] Whewell did not identify the adherents of the 'school of mathematical speculation', but he did demarcate intellectual territory; and by describing his opponents as he did he characterised them as being uninterested in the application of mathematical truth to the uses of society. However, while this may have been true of some of those whom he was opposing, others who were sympathetic to the new Continental methods, such as Babbage and De Morgan, were nevertheless extremely interested in putting mathematics to work.

By the late 1830s there were signs that these debates had forced mathematical speculators to think of themselves more as specialists in pure mathematics than as participants in natural philosophy, which was by then losing its role as the glue that bound the sciences together. In 1838 De Morgan, who had previously expressed enthusiasm for the system of practical technical education offered at the Polytechnic School of Paris,[62] defined the type of mathematical students for whom such an education would be more suitable, and called them 'professional mathematicians':

> ... meaning by that term, 1. The accountant, actuary or mercantile mathematician. 2. The civil engineer, including every species of artificer who uses calculation. 3. The military mathematician. 4. The navigator. 5. The surveyor, including all whose occupation it is to measure quantities. 6. The draughtsman.[63]

These were not, as modern usage of the term 'professional mathematicians' implies, full-time paid mathematical specialists, but quite the opposite: they were members of traditional or newly emerging professions who happened to use mathematics, just the kind of people identified by Wallis and Wallis as constituting the great majority of mathematicians in the eighteenth century.[64] De Morgan desired not only to give them a separate designation, but also a separate mathematical training, one more suited to their needs than that offered at Cambridge, or indeed by himself in London. With such 'professional mathematicians' removed from the system, the unqualified term 'mathematician' would refer only to mathematical specialists, for whom the universities would exclusively cater.

A further sign of the desire that mathematicians had to define better their sphere of operations came in 1837 with the first issue of the *Cambridge Mathematical Journal*. Duncan F Gregory, the editor, said in his Preface that:

> It has been the subject of regret with many persons, that no proper channel existed, either in this University or elsewhere in this country, for the publication of papers on Mathematical subjects, which did not appear to be of sufficient importance to be inserted in the Transactions of any of the Scientific Societies; the two Philosophical Journals which do exist having their pages generally devoted to physical subjects.[65]

The journal was directed at 'all classes of students', and Gregory hoped that it would encourage 'original research'.[66] Such a publication was a new development in British mathematics, and after canvassing for contributions he promised to include 'abstracts of important and interesting papers that have appeared in the memoirs of foreign academies', in order to keep his readers 'on a level with the progressive state of Mathematical Science, and so lead them to feel a greater interest in the study of it'. So now, for the first time, England had a specialist mathematical journal that would allow mathematicians and mathematical students to identify themselves as such by publishing research, and to enjoy a sense of common enterprise with others engaged in the same discipline, both at home and abroad. However, encouraging mathematical research was one thing; providing opportunities by which such work could be made to pay was quite another. The Cambridge mathematical coach William Hopkins wrote in 1841 that:

> Our system, consequently, supplies in itself scarcely any adequate motive for the subsequent extension of our knowledge beyond the bounds prescribed by the general examination. ... The want of encouragement for the ardent and continued prosecution of scientific study within the University, exists also without it in an equal degree ... There are scarcely a dozen public situations in this country to which a high degree of mathematical attainment is an essential recommendation. The only certain inducements for the prosecution of mathematical study are within the University ...[67]

Thus, whilst the specialised study of mathematics may have seemed highly desirable to the editor of the *Cambridge Mathematical Journal*, there was little incentive to take it up.

The 1840s saw increasing public and political disquiet that the ancient universities were failing to use their unparalleled resources in the national interest, and so in 1850 Royal Commissions for Oxford and Cambridge were set up, which both reported in 1852. The Commissioners for Oxford had little to say concerning mathematics because it played such a small part in the life of the university, a situation that they condemned:

> The number of Candidates for Mathematical Honours is very small. As the study of these subjects is, at present, even when most successfully cultivated, almost entirely unproductive of substantial benefit to the Student in securing Scholarships or Fellowships; and as the Professorships of Mathematical Science are so poorly endowed as not to be tenable without other means, it is no wonder that so few Candidates contend for barren honours. ... It ought to be known that there are, or were very lately, Colleges in Oxford where no Mathematical Instruction whatever was supplied to Students.[68]

Nevertheless, the tradition was so well entrenched that not much had changed almost 80 years later when Hardy, then holding the Savilian chair at Oxford, asked:

> Why then should Oxford tolerate its mathematicians but be so totally uninterested in what they do? ... Here Exeter, Lincoln, Magdalen, Oriel, Pembroke, Trinity, University, Wadham, and Worcester [Colleges] have not a mathematician between them; in

none of those great centres of philosophy is there a man who can hope to understand Hilbert or Russell, Einstein or Schrödinger.[69]

The Commissioners for Cambridge, however, elevated the study of mathematics to a matter of national concern, saying that 'On the importance of [mathematical] studies it is impossible to dwell too strongly ... it is impossible not to feel that the glory of the country itself is intimately bound up with our national progress in this direction ...'[70] The Commissioners saw that mathematics had a vital role to play in Britain's industrial and commercial prosperity, and accordingly they were particularly concerned to understand the extent to which mathematical studies were pursued beyond the requirements of the BA degree; in effect, they wanted to know about research. The evidence of Harvey Goodwin, a Cambridge tutor and textbook author, suggested that, in the 10 years since William Hopkins wrote, little had changed:

> The University has no direct means of encouraging the continuance of mathematical reading; the 'Cambridge Philosophical Society,' however, performs an important part in giving a stimulus to original productions, and the inspection of the Transactions of that Society, and also of the volumes of the 'Cambridge and Dublin Mathematical Journal,' will show that at present the University is far from deficient in men who have pursued Mathematics with success. ... the real bar to the pursuit of science seems to me to be ... the difficulty which exists in England of a man living (out of College) by means of his scientific pursuits.[71]

The *Cambridge Mathematical Journal* had amended its name to the *Cambridge and Dublin Mathematical Journal* so as to encourage contributions from Trinity College, Dublin.[72] Thus, while exposure might be given to students' original work by providing opportunities for the reading and publication of papers, a graduate without private resources who did not obtain a fellowship was almost certainly obliged to earn a living by some means other than mathematics.

The idea of making new mathematics more readily available, and of encouraging students to undertake research, was abhorrent to William Whewell. During the 1840s, and together with influential supporters such as Airy – who by then was Astronomer Royal – he attempted to re-establish the traditional basis of mathematics at Cambridge and met with some success, for his appointment in 1841 as Master of Trinity had extended his power and influence. Consequently, his principal aims of stabilising the curriculum and purging it of anything modern or novel were given modest effect in 1846 with a new mathematics syllabus and a Board of Mathematical Studies to supervise the Mathematical Tripos.[73] However, he had been driven to adopt an extreme position, which did nothing to allay the fears of those who saw the ancient universities as failing to make an appropriate contribution to the increasing requirements of commerce, industry and Empire. A letter dated 27 October 1847 from Sir Robert Peel, the former prime minister, to Prince Albert, newly elected as Chancellor of Cambridge University, illustrates the tension between the forces of reaction and the forces of reform, and the extent to which Whewell's views were considered risible in high places:

> I think Dr. Whewell is quite wrong in his position – that mathematical knowledge is entitled to *paramount* consideration, because it is conversant with indisputable truths – that such departments of science as Chemistry are not proper subjects of academical instruction, because there is controversy respecting important facts and principles, and constant accession of information from new discoveries – and danger that the students may lose their reverence for Professors, when they discover that the Professors cannot maintain doctrines as indisputable as mathematical or arithmetical truths.
>
> The Doctor's assumption, that *a century should pass* before new discoveries in science are admitted into the course of academical instruction, exceeds in absurdity anything which the bitterest enemy of University Education would have imputed to its advocates.[74]

Here is not just an extreme point of view but also a claim that mathematics was more than its practical role as servant of the sciences would suggest, because Whewell was presenting it as queen, as 'knowledge ... conversant with indisputable truths', to be compared favourably with the more transient knowledge offered by the new breed of scientists such as chemists.

Thus Oxford and Cambridge; but elsewhere in the 1840s British mathematics was heading in new directions without reference to the ancient universities. Some of the problems caused by developments in algebra have already been described, and in 1830 George Peacock at Cambridge attempted to keep these within bounds by his 'Principle of the Permanence of Equivalent Forms'.[75] In broad terms, this said that an algebraic expression that was valid when the symbols referred to positive integers was also valid when the range of the symbols was widened – even if the symbols then appeared to be meaningless – provided that a consistent interpretation could be found. For example, if n is a positive integer then the symbol a^n represents a multiplied by itself n times, and so has arithmetical meaning, but it has no such meaning if $n = 1/2$. Nevertheless, the principle allowed the use of $a^{1/2}$ in algebra, but also required all exponents, whether positive integers or not, to be combined in the same way; this forced $a^{1/2}a^{1/2} = a^{1/2+1/2} = a^1 = a$, from which it followed that $a^{1/2}$ must be interpreted as $\sqrt{a}$.[76] Extending the argument, $a^{1/n}$ (where n is a positive integer) was interpreted as the nth root of a. Likewise, a^0 was meaningless, but in this case the requirement was that $a^n a^0 = a^{n+0} = a^n$ for any positive integer n, from which it followed that a^0 must be interpreted as 1. The principle then allowed use of the unintuitive a^{-n}, because of the requirement that $a^n a^{-n} = a^{n-n} = a^0 = 1$, which forced the interpretation of a^{-n} as $1/a^n$. In all these cases it was the demand for consistency of treatment and an absence of contradiction that permitted particular interpretations to be placed upon otherwise meaningless algebraic symbolism. Of course, exponents had long been used in this way, but Peacock had now provided an intuitively reasonable means of determining what was an acceptable extension of algebra, and what was beyond the pale.

Peacock's principle was widely adopted so as to legitimise an extended range for algebra, which eventually resulted in a point of view that was neatly encapsulated by William Spottiswoode when referring to the equation $x^2 + 1 = 0$:

> What meaning can be attached to a root of this equation in the special applications of Algebra to Geometry or Physics, or whether any meaning at all can be given to it, is, after all, a question of only secondary importance to the pure analyst. To him, the main question is, In what relation does an expression capable of serving as a root of this equation stand to other and previously known expressions of Algebra?[77]

However, for some mathematicians Peacock's principle was disturbing because it allowed the use of symbolism that had no obvious, natural meaning. Instead it involved interpretations, and it was not clear how, if at all, interpretations based on a principle could reveal the eternal, immutable truths that were supposed to lie at the heart of mathematics.

Nevertheless, the principle had a long life, as Bertrand Russell discovered when, as a mathematics student at Cambridge in the 1890s, he raised a question concerning real numbers:[e]

> Russell still felt resentment against his teachers in 1960. In conversation he recalled that though they only proved the binomial [theorem] for a positive integral exponent, they stated and used it with a real exponent. 'I asked them how the general case could be proved. "We prove it" they replied, "by the principle of permanence of form"!'[78]

Peacock's principle also had the consequence that algebra had to be commutative, because commutativity – the rule that $a + b = b + a$ and $ab = ba$ – expresses a basic property of the addition and multiplication of positive integers. Therefore it caused consternation when, during the 1840s and well away from Cambridge, there emerged the non-commutative algebra of quaternions, which was invented in Dublin by the Irish mathematician William Rowan Hamilton. As well as the real numbers it used three symbols i, j and k which were postulated as having the property that $i^2 = j^2 = k^2 = ijk = -1$; with these equations and the associative law of arithmetic, one can show that $ij \neq ji$.[f] Quaternions were soon to be of value in mathematical physics, and something of the awe and wonder that they inspired in contemporary mathematicians can be gleaned from William Spottiswoode:

> Well do I remember how in its early days we used to handle the method as a magician's page might try to wield his master's wand, trembling as it were between hope and fear, and hardly knowing whether to trust our own results until they had been submitted to the present and ever-ready counsel of Sir W. R. Hamilton himself.[79]

There was also the non-commutative algebra of matrices, which soon became of considerable interest and use in many departments of

---

[e] Real numbers comprise all the numbers that are used in elementary arithmetic, namely the positive and negative integers (including zero), fractions, and infinite non-recurring decimals (finite and recurring decimals can be expressed as fractions).

[f] The associative law says that in the addition or multiplication of three terms, $(a + b) + c = a + (b + c)$, and $(ab)c = a(bc)$. It is this law that enables one to write $a + b + c$ and abc without ambiguity. So, using the defining equations for quaternions, we have: $i^2 = ijk$, so $i = jk$, and $ijk = k^2$, so $ij = k$; but then $ji = jjk = j^2k = (-1)k = -k$, and so $ij \neq ji$.

mathematics; it was mainly the brainchild of Arthur Cayley, the senior wrangler and first Smith's prizeman of 1842. A matrix is a rectangular array – of numbers, in its simplest form – that is subject to rules by which, in certain circumstances, two matrices can be combined to produce a third, using methods conventionally referred to as addition and multiplication. However, the rule for multiplication is such that the order in which the matrices are combined can affect the outcome: thus $\begin{pmatrix} 1 & 2 \\ 3 & 4 \end{pmatrix} \begin{pmatrix} 5 & 6 \\ 7 & 8 \end{pmatrix} = \begin{pmatrix} 19 & 22 \\ 43 & 50 \end{pmatrix}$, but $\begin{pmatrix} 5 & 6 \\ 7 & 8 \end{pmatrix} \begin{pmatrix} 1 & 2 \\ 3 & 4 \end{pmatrix} = \begin{pmatrix} 23 & 34 \\ 31 & 46 \end{pmatrix}$; the two multiplied matrices do not commute. As with quaternions, this development took place away from Cambridge, because at the time Cayley was working as a conveyancing barrister in London; although there is little doubt that he could have obtained a permanent position at Cambridge after taking the Tripos, he probably would have had to take holy orders, and this he was unwilling to do. Instead, he turned to the law, which remained his profession until 1863, when Cambridge instituted its first chair in pure mathematics, the Sadleirian. As holy orders were not in general a requirement for professors, Cayley – by then widely recognised as Britain's leading pure mathematician – was invited to take up the chair, which he occupied until his death more than 30 years later.[80]

Just as quaternions and matrices were gaining attention, there appeared in 1847 *The Mathematical Analysis of Logic: Being an Essay towards a Calculus of Deductive Reasoning* by George Boole. He was the headmaster of a small school in Lincoln, had no university degree, and held no university appointment until in 1849 he accepted the chair of mathematics at Queen's College, Cork. In 1854 he developed his ideas further in what became a classic work entitled *An Investigation of the Laws of Thought*, which introduced an algebraic system for dealing with the truth and falsity of logical propositions. This system later became known as Boolean algebra, and at the beginning of the twentieth century Bertrand Russell wrote: 'Pure mathematics was discovered by Boole, in a work which he called the *Laws of Thought*. This work abounds in asseverations that it is not mathematical, the fact being that Boole was too modest to suppose his book the first ever written on mathematics.'[81] Russell was here giving exaggerated expression to his belief that the foundations of mathematics should be formulated using the principles of symbolic logic that had been developed during the second half of the nineteenth century, and of which he considered

Boole to be the progenitor. The upshot was that Russell co-authored with Alfred North Whitehead the three-volume *Principia Mathematica*, publication of which began in 1910. Boolean algebra has since become of great significance in computer science.

Closely connected to the work of Boole was Augustus De Morgan's formulation in London of some principles for symbolic logic that are still referred to as De Morgan's laws. In these, he stated two relations between propositions that themselves included references to other propositions p and q. The first law says that '"it is false that p or q is true" is true if, and only if, "p is false and q is false" is true'; and the second law says that '"it is false that p and q are true" is true if, and only if, "p is false or q is false" is true'; these laws are fortunately much more concise when expressed symbolically. De Morgan was one of the best-known Victorian mathematicians; he had been fourth wrangler in 1827, and in 1828 was appointed London University's first professor of mathematics at the age of 22. After three years he resigned on a point of principle, but in 1836 was reappointed at University College, a position that he was to retain until in 1866 he resigned for a second time, again on a point of principle, although he continued to teach until the session ended the following year.[82] He possessed a particularly logical, pedantic and scrupulous cast of mind that rejected many of the Cambridge ideals and embraced instead those of the University of London. He was a very able mathematician who had assimilated Continental mathematics with enthusiasm and precision; he championed the founding of the calculus on the idea of the limit; and his interest in logic and the processes of reasoning enabled him to appreciate, and offer influential opinions on, many of the difficult issues that new developments in the calculus and algebra were generating. Unlike the professors at Cambridge, he believed that his principal function was to teach, a task to which he was dedicated and in which he was particularly effective. He was therefore extremely visible to his students, and his opinions concerning the nature of mathematics were disseminated far more widely than were those of any contemporary British mathematician. His influence was greatly increased by the extent of his writing for the general reader, particularly on educational matters, and he contributed more than 800 articles to the popular *Penny Cyclopaedia*.[83]

These developments show that during the first half of the nineteenth century some mathematicians became increasingly involved with mathematics that had only a tenuous connection with traditional natural philosophy. The extent to which pure mathematics was displaying ever-increasing abstraction suggests that it could well have shared the fate of another discipline with ideals that were incompatible with the dominant scientism of the Victorian era, namely the classics. Until the end of the nineteenth century the teaching of classics was the mainstay of British public schools and grammar schools, many of which had been founded for that very purpose. Consequently classics teachers enjoyed pay, conditions and status superior to those of their colleagues in less well-regarded disciplines such as mathematics; and at some schools, even as late as the mid-Victorian period, the mathematics master was 'not allowed to wear a gown or to have any part in the discipline of the school'.[84] As for the universities, classics and theology had always been the principal disciplines at Oxford, but at Cambridge they were eclipsed by mathematics during the eighteenth century. However, after the introduction of the Classical Tripos in 1824, and particularly after the abolition in 1851 of the requirement imposed on all candidates (except the sons of peers) to first have succeeded in the Mathematical Tripos, the classics staged a revival; and by 1863 the notion that at Cambridge it was not as well-regarded as mathematics was thought to be so damaging that a strong refutation was required. This was duly provided in that year by the historian John Robert Seeley, who wrote in *The Student's Guide to the University of Cambridge*:

> It used to be the received opinion, and for a long time it was a just opinion, that classical studies were little pursued or valued at Cambridge. That this has now entirely ceased to be true, is well known to all who understand the present condition of the Universities; but such persons are not numerous, and to the majority, I think, the assertion will still be novel and difficult to believe. Let the incredulous then observe how little shorter the Classical Honour List is than the Mathematical; let them take notice of the University Scholarships which are annually given for Classics and contended for generally by seventy or eighty men, and of the numerous prizes annually given for compositions of various kinds in Latin and Greek, rewards far outnumbering those offered for Mathematical proficiency; let them also remember that no precedence is now given to Mathematics in any one point; and they will perhaps be convinced of the fact, that classical studies are now equally

esteemed and not much less practised at Cambridge than mathematical. But they may even then be unprepared for this further assertion, which notwithstanding is made with confidence, that as a place of Classical scholarship and training, Cambridge is fully equal to Oxford.[85]

This revival of fortunes points to a renewed enthusiasm amongst Cambridge students for immersing themselves in the language and literature of ancient Greece and Rome, yet the everyday problems for which the Odes of Horace would supply material assistance must have been few and far between; rather, it was believed that a mind that had been trained in that fashion had at its command a resource of incalculable benefit that would ensure precision of thought and argument in any walk of life.

This was just the kind of educational advantage that was also claimed for mathematics, and pure, abstract mathematics provided an even closer parallel, in that it did not encompass those aspects of the discipline that directly addressed empirical questions. At Cambridge, Thomas Worsley of Downing College expressed it thus in 1865, when he allowed geometry to stand proxy for mathematics because of the great esteem in which Euclidean geometry was then held:

> We have seen that the grammar and logic of language, exemplified for us and wrought into us by our accurate study of Greek and Latin, educe the fundamental conceptions and forms of thought which language involves, better than can any mere reading, however extensive, of the books written in these languages. And, in like manner, the abstract truths of geometry and mechanics, exemplified for us and wrought into us by our primary scientific discipline, educe the fundamental conceptions and forms of thought which science involves, more effectually than can any mere cumulative knowledge of the facts of the physical universe, or of mathematical expressions for them.[86]

Worsley took a tilt at Oxford in saying that the study of the 'grammar and logic of language' was superior to the 'mere reading' of books, for this encapsulated the different approaches to classical pedagogy that obtained at the two universities. The emphasis at Cambridge on grammar and logic, and so on accurate translation, allowed the setting of examination questions that could be precisely marked, thus making feasible an order of merit in the Classical Tripos, just as in the Mathematical.[87] Therefore at Cambridge both the classics and mathematics were offered not only as training for the mind but also as a means of providing the competition between students that the university so prized, but that was foreign to Oxford pedagogy. Seeley summed it up: 'It should most decidedly be understood that persons who wish to avoid competition, whether in Classics or Mathematics, had better not come to Cambridge.'[88] In both disciplines, regarded as pedagogical tools, Cambridge demanded precise thought, accurate work and certain results, and this was reflected by the syllabuses: Cambridge's approach to the classics was paralleled in mathematics by a reluctance to introduce new topics that might call into question the certainty of mathematical demonstration.

Furthermore, the lack of career prospects hampered both classical and mathematical scholars alike, as Richard Sheepshanks wrote in 1854 – he too thought of mathematics as geometry:

> Up to a certain point, and for certain subjects, our system of grammar-schools, colleges, and universities, is only faulty in details, and admits of easy, gradual, and indefinite improvement. But there is no subsequent career; and our powerful geometers and splendid classics look naturally for some mode of living which the professions alone offer. A few men, who feel their call strongly, and who *will* follow the bent of their own talents ... are produced occasionally, and upon these England chiefly depends for her exact sciences. ... One of our best mathematicians, Sir John Lubbock, and one of our most learned scholars, Mr. Grote, are found, where one would scarcely look for them, among the leading bankers of London.[89]

The reference to 'Mr. Grote' was particularly pertinent because George Grote was at that time just completing his monumental *History of Greece*. Given these parallels, it is reasonable to suppose that the fate of the two sets of practitioners – classical scholars and pure mathematicians – would have been similar in the face of the pressure that they were both under from the newly emerging scientific specialists, yet what occurred was very different. The resurgence of interest in the classics did not last, and for more than a century its proponents have been fighting a rearguard action to preserve the significance and status of their steadily declining discipline, whereas pure mathematicians have found no need to do likewise. Christopher Stray has described how 'the high-cultural claims of Victorian Hellenism gave way early in the twentieth century to a lower-key disciplinary formation whose symbolic centre of gravity was Latin rather than Greek'. However, the reign of Latin was short-lived: 'The universities' abandonment of compulsory Latin [in the mid-twentieth-century] marks a crucial stage in the marginalizing of classics in whatever form.'[90] Lack of usefulness and irrelevance to contemporary concerns – characteristics that have often been used to castigate classicists – applied equally to pure mathematicians, yet in spite of this they were eventually able to secure an entitlement to pursue their vocation equal to that enjoyed by practitioners in the applied sciences.

In the following chapters we shall investigate the acquisition of this entitlement by pure mathematicians in Victorian Britain. So far we have seen that by the 1850s the long-standing pre-eminence of Cambridge University as the focus of mathematical activity in England was apparently threatened. The university had a well-established tradition in which mathematics was not only the servant of Newtonian natural philosophy, but also the best way of inculcating in students modes of thought and powers of reasoning that were deemed to be invaluable in any discipline; and supporters of this tradition were attempting to stand firm against a competing vision, espoused by the new breed of pure mathematicians, in which mathematics was principally studied for its own sake rather than its utility, and in which abstraction was a virtue, not a defect. Mathematicians based outside Cambridge had a virtual monopoly over whatever new pure mathematics Britain produced, and three of the most influential – Arthur Cayley, James Joseph Sylvester and Augustus De Morgan – all lived and worked in London; so it is to events there that we shall turn next.

Notes and References

1. See in particular Dee's 'Groundplat' (or 'Groundplatt') in *The Elements of Geometrie of the most auncient Philosopher EVCLIDE of Megara. Faithfully (now first) translated into the Englishe toung, by H. Billingsley, Citizen of London ... With a very fruitfull Præface made by M. I. Dee, specifying the chiefe Mathematicall Scieces, what they are, and wherunto commodious: where, also, are disclosed certaine new Secrets Mathematicall and Mechanicall, vntill these our daies, greatly missed* (London: John Daye, 1570).
2. John Wallis, 'Dr. Wallis's Account of Some Passages of his Own Life', in Thomas Hearne, *Peter Langtoft's Chronicle, (as illustrated and improv'd by Robert of*

Brunne) from the Death of Cadwaladar to the End of K. Edward the First's Reign. To which are added ... other curious Papers ... (Oxford: Printed at the Theater, 1725), pp. cxl–clxx; the manuscript transcribed by Hearne is dated 29 January 1696/7.
3. Isaac Barrow, *The Usefulness of Mathematical Learning explained and demonstrated: being Mathematical Lectures read in the Publick Schools at the University of Cambridge, to which is prefixed, the Oratorical Preface of our Learned Author, spoke before the University on his being elected Lucasian Professor of the Mathematics*, translated by the Rev. Mr John Kirkby, of Egremond in Cumberland (London: Stephen Austen, 1734), p. 50; the Latin text of Barrow's preface and lectures, delivered in 1664, was published posthumously in 1683.
4. Thomas L Heath, *A Manual of Greek Mathematics* (Oxford: Clarendon Press, 1931), p. 6.
5. Francis Bacon, *The Twoo Bookes of Francis Bacon, Of the Proficience and Advancement of Learning, Divine and Humane* (At London, Printed for Henrie Tomes, and are to be sould at his shop at Graies Inne Gate in Holborne, 1605), p. 31.
6. Narcissus Luttrell, *A Brief Historical Relation of State Affairs, 1678–1714*, 1 (Oxford: Oxford University Press, 1857), p. 396; this records a diary entry for March 1686/7.
7. Samuel Johnson, *A Dictionary of the English Language: In which the Words are Deduced from their Originals, and illustrated in their Different Significations by Examples from the Best Writers. To which are prefixed, a History of the Language, and an English Grammar* (London: J & P Knapton, 1755), Art. 'Mathematicks'.
8. For the use of these terms see, for example, Barrow, Usefulness, p. 11; Thomas Walter, *A new Mathematical Dictionary, containing the Explication of all the Terms in Pure and Mixed Mathematics ... To which is prefixed, the Elements of Geometry* (London: The Author, 1761), Art. 'Mathematics'; Peter Barlow, *A new Mathematical and Philosophical Dictionary; comprising an Explanation of the Terms and Principles of Pure and Mixed Mathematics ... And an Account of the Discoveries and Writings of the most celebrated Authors, both ancient and modern* (London: G & S Robinson, 1814), Art. 'Mathematics'.
9. Barrow, *Usefulness*, pp. 50–51.
10. Walter, *Mathematical Dictionary*, Art. 'Mathematics'.
11. R V Wallis and P J Wallis, *Index of British Mathematicians, Part III: 1701–1800* (Newcastle upon Tyne: Project for Historical Biobibliography, 1993), pp. viii, x.
12. [Anon.] Peter Shaw, *Man. A Paper for Ennobling the Species*, 35 (27 August 1755), p. 3. This weekly periodical was issued anonymously, but authorship is attributed to the physician Peter Shaw in the English Short Title Catalogue.
13. Sophia De Morgan, *Memoir of Augustus De Morgan* (London: Longmans, Green, 1882), p. 305.
14. *Clarendon Report*, I (London: Eyre & Spottiswoode, 1864), pp. 214 (Harrow), 81 (Eton); for the teaching of elementary arithmetic at Eton, see the evidence of Rev. S T Hawtrey (III, p. 217, questions 6261–6287). [Reports of the Commissioners, vol. 20.]
15. Andrew Warwick, *Masters of Theory: Cambridge and the Rise of Mathematical Physics* (Chicago: University of Chicago Press, 2003), p. 34.
16. Quoted in James D Forbes, *A Review of the Progress of Mathematical and Physical Science in more Recent Times, and particularly between the Years 1775 and 1850; being one of the Dissertations prefixed to the Eighth Edition of the Encyclopaedia Britannica* (Edinburgh: A & C Black, 1858), p. 8, par. 29.
17. Letters in Sophia De Morgan, *Memoir*, pp. 388, 390.

18. Philip C Enros, 'The Analytical Society (1812–1813): Precursor of the Renewal of Cambridge Mathematics', *Historia Mathematica*, 10 (1983), 24–47, at 29.
19. Wilfrid Airy (ed.), *Autobiography of Sir George Biddle Airy, K.C.B., M.A., LL.D., D.C.L., F.R.S., F.R.A.S., Honorary Fellow of Trinity College, Cambridge, Astronomer Royal from 1836 to 1881* (Cambridge: Cambridge University Press, 1896), p. 24.
20. Warwick, *Masters of Theory*, pp. 77, 144–151.
21. J M F Wright, *A Supplement to Wood's Algebra; as given in the Private Tutor* (Cambridge: W P Grant, 1831), p. 45.
22. Walter, *Mathematical Dictionary*, Art. 'Fluxions'.
23. Pierre Duhem (tr. Philip P Weiner), *The Aim and Structure of Physical Theory*, second edition (New York: Atheneum, 1974), pt. 1, ch. 4; first French edition 1906.
24. S F Lacroix, *An Elementary Treatise on the Differential and Integral Calculus* (Cambridge: Deighton, 1816), p. 596, n. B (by Peacock). Translated from the French by Babbage, Peacock and Herschel; the appendix is an original treatise by Herschel.
25. Augustus De Morgan, *The Differential and Integral Calculus* (London: Baldwin & Cradock, 1842), p. v.
26. Augustus De Morgan, 'On Fractions of Vanishing or Infinite Terms', *The Quarterly Journal of Pure and Applied Mathematics*, 1 (1855), 204–210, at 204.
27. J R Tanner (ed.), *The Historical Register of the University of Cambridge, being a Supplement to the Calendar with a Record of University Offices, Honours and Distinctions to the Year 1910* (Cambridge: Cambridge University Press, 1917), p. 602.
28. J M Dubbey, *The Mathematical Work of Charles Babbage* (Cambridge: Cambridge University Press, 1978), pp. 24–25.
29. Dubbey, *Babbage*, p. 25.
30. For a further argument that the Analytical Society had little effect on British mathematics, see Kaila Katz, *The Impact of the Analytical Society on Mathematics in England in the First Half of the Nineteenth Century*, being a PhD thesis (1982) for New York University (Ann Arbour, MI: University Microfilms International, 1985).
31. Tanner, *Historical Register*, p. 83, n. 10; Dubbey, *Babbage*, p. 7.
32. R Sheepshanks, *A Letter to the Board of Visitors of the Greenwich Royal Observatory in reply to the Calumnies of Mr. Babbage at their Meeting in June 1853, and in his Book entitled* The Exposition of 1851 (London: G Barclay, 1854), p. 78.
33. Matthew Arnold, *Schools and Universities on the Continent* (London: Macmillan, 1868), chs. VIII (France) and XX (Prussia); see also Gert Schubring, 'Germany to 1933', in Ivor Grattan-Guinness (ed.), *Companion Encyclopedia of the History and Philosophy of the Mathematical Sciences* 2 (London: Routledge, 1994), pp. 1442–1456.
34. R Steven Turner, 'The Growth of Professorial Research in Prussia, 1818–1848: Causes and Context', in Russell McCormmach (ed.), *Historical Studies in the Physical Sciences* (Philadelphia: University of Pennsylvania Press, 1971), pp. 137–182.
35. Charles Babbage, *Reflections on the Decline of Science in England, and on some of its Consequences* (London: Fellowes, 1830), ch. II.
36. OED Online, Art. 'Applied, adj.'
37. Barlow, *Dictionary*, p. v.
38. Charles Hutton, *A Philosophical and Mathematical Dictionary: containing an Explanation of the Terms, and an Account of the several Subjects, comprised*

under the heads Mathematics, Astronomy, and Philosophy both Natural and Experimental; with an Historical Account of the Rise, Progress and Present State of these Sciences; also Memoirs of the Lives and Writings of the most eminent Authors, both ancient and modern, who by their Discoveries or Improvements have contributed to the Advance of them (London: The Author, 1815), I, pp. 60–61. This is the second edition, and Hutton explained algebra in the same terms as he had used in the first edition of 1796 (I, pp. 63–64).

39. George Peacock, *Collection of Examples of the Applications of the Differential and Integral Calculus* (Cambridge: Deighton, 1820), p. iii.
40. John Playfair, 'Dissertation Second: Exhibiting a General View of the Progress of Mathematical and Physical Science, since the Revival of Letters in Europe', in *Supplement to the Fourth, Fifth, and Sixth Editions of the Encyclopaedia Britannica* 2 (Edinburgh: Archibald Constable, 1824), p. 1.
41. For contemporary views on these matters, see *The London Encyclopaedia, or Universal Dictionary of Science, Art, Literature, and Practical Mechanics, comprising a Popular View of the Present State of Knowledge. By the Original Editor of the Encyclopaedia Metropolitana, assisted by eminent Professional and other Gentlemen* (London: Thomas Tegg, 1826–1829), XIII, Art. 'Mathematics', pp. 677–678. The 'Original Editor' was Thomas Curtis.
42. F J C Hearnshaw, *The Centenary History of King's College London: 1828–1928* (London: Harrap, 1929), p. 68.
43. J R Young, *An Introductory Lecture, delivered at the opening of the Mathematical Classes of Belfast College, November 12, 1833* (London: J Souter, 1833), p. 30.
44. *Report of Her Majesty's Commissioners Appointed to Inquire into the State, Discipline, Studies, and Revenues of the University and Colleges of Cambridge: Together with the Evidence, and an Appendix* (London: Her Majesty's Stationery Office, 1852), Evidence, p. 229. [Reports of the Commissioners, vol. 44.]
45. Johnson, *Dictionary*, Art. 'Algebra'.
46. Nathan Bailey, *An Universal Etymological English Dictionary*, twentieth edition (London: T Osborne, 1763), Art. 'Algebra Numeral or Vulgar'.
47. Bailey, *Dictionary*, Art. 'Algebra Literal or Specious'.
48. Walter, *Mathematical Dictionary*, Art. 'Algebra'.
49. Barlow, *Dictionary*, Art. 'Algebra'.
50. Archibald Sandeman, *Pelicotetics or the Science of Quantity: An Elementary Treatise on Algebra and its Groundwork Arithmetic* (Cambridge: Deighton Bell, 1868), pp. viii–x.
51. De Morgan, *Memoir*, p. 19.
52. *Encyclopaedia Britannica; or, a Dictionary of Arts, Sciences, and Miscellaneous Literature*, enlarged and improved, sixth edition, I (Edinburgh: Archibald Constable, 1823), Art. 'Algebra', p. 605.
53. See his letter to John Herschel dated 6 March 1817, quoted in Dubbey, *Babbage*, p. 41.
54. William Whewell, *Thoughts on the Study of Mathematics, as Part of a Liberal Education*, Second Edition (Cambridge: Deighton, 1836), p. 11. This work was also included as an appendix in his *On the Principles of an English University Education* (London: Parker, 1837).
55. Whewell, *Thoughts*, p. 5.
56. Whewell, *Thoughts*, pp. 5–6.
57. Whewell, *Thoughts*, p. 13.
58. Whewell, *Thoughts*, pp. 12–13, 18–19.
59. *Cambridge Commissioners*, Evidence, p. 272.

60. Whewell, *Thoughts*, p. 35.
61. Whewell, *Thoughts*, p. 41.
62. Augustus De Morgan, 'Polytechnic School of Paris', *Quarterly Journal of Education*, 1(1) (January–April 1831), 57–74. The article is anonymous, but the attribution is in Sophia De Morgan's Memoir of her husband, p. 401.
63. Augustus De Morgan, 'Professional Mathematics', in *Central Society of Education, Second Publication* (London: Taylor & Walton, 1838), p. 133.
64. Wallis and Wallis, *Index*.
65. [Anon.] Duncan Gregory, 'Preface', *Cambridge Mathematical Journal*, I (November 1837), 1–2, at 1. The first volume of the CMJ was issued in 1839 and comprised numbers spanning two years; the unsigned Preface appeared in the first number. See Sloan Evans Despeaux, '"Very Full of Symbols": Duncan F. Gregory, the Calculus of Operations, and the *Cambridge Mathematical Journal*', in Jeremy Gray and Karen Hunger Parshall (eds.), *Episodes in the History of Modern Algebra (1800–1950)* (Providence, RI: American Mathematical Society and The London Mathematical Society, 2007), pp. 51, 70; also Tony Crilly, 'The Cambridge Mathematical Journal and its Descendants: The Linchpin of a Research Community in the Early and Mid-Victorian Age', *Historia Mathematica*, 31 (2004), 455–497.
66. Sloan Evans Despeaux, 'Launching Mathematical Research without a Formal Mandate: The Role of University-Affiliated Journals in Britain', *Historia Mathematica*, 34 (2007), 89–103, at 96–97.
67. W Hopkins, *Remarks on certain proposed Regulations respecting the Studies of the University and the Period of conferring the Degree of B.A.* (Cambridge: Deighton, 1841), pp. 12–13.
68. *Report of Her Majesty's Commissioners appointed to inquire into the State, Discipline, Studies, and Revenues of the University and Colleges of Oxford: Together with the Evidence, and an Appendix* (London: Her Majesty's Stationery Office, 1852), pp. 63–64. [Reports of the Commissioners vol. 22.]
69. G H Hardy, 'Mathematics', *The Oxford Magazine*, 48 (5 June 1930), 819–821, at 819, 820.
70. *Cambridge Commissioners*, Report, p. 105.
71. *Cambridge Commissioners*, Evidence, p. 237.
72. Despeaux, 'Launching Mathematical Research', 98.
73. Harvey W Becher, 'Radicals, Whigs and Conservatives: The Middle and Lower Classes in the Analytical Revolution at Cambridge in the Age of Aristocracy', *British Journal for the History of Science*, 28 (1995), 405–426, at 425.
74. Quoted in Theodore Martin, *The Life of his Royal Highness the Prince Consort*, II (London: Smith Elder, 1876), pp. 117–118.
75. George Peacock, 'Preface', *Treatise on Algebra* (Cambridge: Deighton, 1830).
76. Dubbey, *Babbage*, p. 102.
77. William Spottiswoode, 'Remarks on some Recent Generalizations of Algebra', *Proceedings of the London Mathematical Society*, IV (1871), 147–164, at 148.
78. R O Gandy, 'Bertrand Russell, as a Mathematician', *Bulletin of the London Mathematical Society*, 5 (1973), 342–348, at 347.
79. William Spottiswoode, 'Address', in *Report of the Forty-eighth Meeting of the British Association for the Advancement of Science* [1878] (London: John Murray,1879), p. 17.
80. The standard biography of Cayley is Tony Crilly, *Arthur Cayley: Mathematician Laureate of the Victorian Age* (Baltimore, MD: Johns Hopkins, 2006).

81. Bertrand Russell, 'Mathematics and the Metaphysicians', in *Mysticism and Logic and Other Essays* (London: Longmans, 1918), p. 74; originally published 1901.
82. See De Morgan's letter to Sir John Herschel dated 25 March 1867, in De Morgan, *Memoir*, p. 369.
83. See the full list in De Morgan, *Memoir*, pp. 407–414.
84. J R de S Honey, *Tom Brown's Universe: The Development of the Victorian Public School* (London: Millington, 1977), p. 136.
85. *The Student's Guide to the University of Cambridge* (Cambridge: Deighton, Bell, 1863), p. 17.
86. T Worsley, *Christian Drift of Cambridge Work: Eight Lectures recently delivered in Chapel on the Christian Bearings of Classics Mathematics Medicine and Law Studies prescribed in its Charter to Downing College* (London: Macmillan, 1865), p. 55.
87. Joseph Williams Blakesley, *Where Does the Evil Lie? Observations Addressed to the Resident Members of the Senate on the Prevalence of Private Tuition in the University of Cambridge* (London: Fellowes, 1845), p. 38.
88. *Student's Guide*, p. 18.
89. Sheepshanks, *Letter*, p. 76.
90. Christopher Stray, *Classics Transformed: Schools, Universities, and Society in England, 1830–1960* (Oxford: Clarendon Press, 1998), p. 1.

3 THE LONDON MATHEMATICAL SOCIETY

In 1868 Professor Leone Levi, the economist and jurist, presented to the Norwich meeting of the British Association for the Advancement of Science a paper entitled 'On the Progress of Learned Societies, illustrative of the Advancement of Science in the United Kingdom during the last Thirty Years'. After what must have been a somewhat numbing address, during which he 'dwelt at length on the circumstances attending almost every scientific society in the United Kingdom', he summarised, with statistical tables that underpinned his analysis, the results of an appeal for information that he had made to the societies, of which he identified 73.[1] The great majority had indeed been founded during the preceding 30 years, and they catered for almost every scientific interest; their aggregate membership of 57,127 reflected a very characteristic Victorian enthusiasm.

The advantages that these new societies gave their members were not confined to providing a means of communicating or socialising with fellow enthusiasts, or giving them a sense of confraternity when engaged in what might otherwise have been disparate and disorganised activities; a society also allowed its members to be publicly recognised as a group with particular skills, or access to knowledge of a specialised kind. Some of these societies aimed for the professionalisation of their discipline, so as to raise the social status of their members and provide them with greater career rewards, in return for which they would – in theory, at least – act for the public good rather than private gain.[a] Other

---
[a] The topic of professionalisation is fully dealt with in Chapter 6.

societies resulted from the trend towards scientific specialisation that we noted in the previous chapter; in these, the members were presented as custodians and disseminators of knowledge, which would give them an elevated public profile, allow them to influence public and official opinion, and provide a dedicated outlet for their writings.

Levi divided the societies into several categories, and that of mathematical and physical sciences was not well represented when compared with others such as applied sciences, or biology and natural history, both of which had many more members and a much greater income. These more popular and mature societies reflected not only the increasing importance that was attached to specialist technical skills, but also the contemporary preoccupation with the investigation, description, classification, manufacture and use of *things*. In particular, Levi identified only one society dealing explicitly with mathematics, and that was the newest of all, less than four years old and having just 111 members: the London Mathematical Society (LMS).

At this time pure mathematicians, newly arrived on the Victorian mathematical scene, were finding that they needed to justify their activities, because their principles, unlike those of the natural philosophers, rejected not only the natural world and natural quantities as the touchstones of mathematical truth, but also the characterisation of pure mathematics as the servant of the sciences. Consequently, they appeared to be increasingly irrelevant to the concerns of most Victorians, and were left with the question as to whether there was any point to pure mathematics at all, to which they had at first no satisfactory reply. However, a mathematical society, if sufficiently prescient and forward-looking, should have been able to take the lead in attempting to rebut the inevitable criticism, both by devising positive arguments in support of pure mathematical research, and by encouraging British pure mathematicians to participate actively in the new approaches to mathematics that were being developed on the Continent. Fulfilling this brief would have entailed the forging of international avenues of communication such as already existed for applied mathematicians and mathematical physicists; and the LMS, which was founded just as these issues were becoming more prominent, seems to have been ideally placed for such a purpose. In this chapter we shall investigate the extent to which it succeeded in carrying it out.

Levi's analysis might suggest that until then Britain had been a mathematical wasteland, but that is far from being true because, unlike the situation in most other countries, there was a continuing tradition of mathematical involvement by those with little formal education in the subject. This tradition, going back to the early eighteenth century, was manifested by numerous journals that were devoted, in whole or in part, to mathematical problems, puzzles and diversions. Interest in such mathematics was found in all classes of society, and the journals had a wide circulation; their content was by no means trivial, and occasionally they provided material for university examiners. One problem, popularly known as Kirkman's schoolgirl problem and published in *The Lady's and Gentleman's Diary* for 1850, even served to introduce to the public the study of a new branch of mathematics, that of permutations and combinations.[2]

This popular interest in mathematics also generated some local mathematical clubs and societies, although they were not as numerous as the journals. Sometimes the members of these societies were from a trade or occupation for which additional mathematical skill would have been an advantage, but equally often the members just gained intellectual satisfaction from increasing their knowledge and attempting to solve mathematical problems. In *A Budget of Paradoxes* Augustus De Morgan contrasted the situation in Britain with that in Continental Europe, where mathematical education was highly centralised and controlled; this resulted, he said, in some great mathematicians producing profound results, but with no dissemination of mathematical knowledge amongst the lower classes. Referring to Britain, he added that 'In no other country has the weaver at his loom bent over the Principia of Newton; in no other country has the man of weekly wages maintained his own scientific periodical.'[3]

De Morgan's reference to 'the weaver at his loom' was not random, for it was in this occupation that refugees from the Continent often specialised, and the French Huguenots, in particular, attached great importance to self-education. It was probably for this reason that one of the best known mathematical societies arose and flourished in the Spitalfields area of London, just to the east of the present-day Liverpool Street Station, where there was a sizable population of French immigrants; the society became known as the Spitalfields Mathematical Society. It was founded in 1717, and in its early years provided improving lectures on mathematics and allied topics to seamen, artisans and tradesmen, but it then broadened its scope, so that a few decades later it was providing public lectures of general scientific interest. In the nineteenth century it acquired the character of a mathematical club, partly at least because more specialised scientific societies had taken over some of its previous functions, but by 1845 it had completely run out of steam and had to be wound up, with most of the few remaining members being accepted as Fellows by the Royal Astronomical Society, which also took over the library.

De Morgan was a keen member of the Astronomical Society and on a committee that inspected the members of the Spitalfields – who were generally believed not to be well-educated or of good social standing – to ensure that they were sufficiently gentlemanly and knowledgeable to mingle with the members of the Astronomical; the committee was pleasantly surprised to find that its preconceptions were misplaced.[4] Nevertheless, although the members of the Spitalfields passed muster, it had not been a learned society suitable for the presentation of original research. Only the advent of the LMS, nearly 20 years later, gave British pure mathematicians the opportunities that had long been enjoyed by members of other societies such as the Astronomical and Statistical.

If the LMS had been, as its name implies, simply an association of some mathematical enthusiasts in London, then it would probably have had little influence. However, the members were not so limited, and the membership list for the period until the turn of the century includes the names of nearly all the best-known and most prolific mathematicians in Britain. Furthermore, data from the comprehensive *Jahrbuch über die Fortschritte der Mathematik*, which each year listed, classified and abstracted mathematical papers from about 150 journals in many languages, show that a high proportion of the pure mathematical output from British authors came from members of the Society.

In using the *Jahrbuch* as evidence, we have taken comparatively brief papers, rather than books, as the measure of pure mathematicians' contributions to scholarship, and this perhaps needs some explanation. In pure mathematics it is possible to present ideas of great significance within a small compass: there are no empirical results to which to appeal, no analogies to be set up between the theoretical apparatus and the natural world, and mathematical orthography and reasoning are very compressed. Furthermore, a new mathematical result is frequently significant in its own right, and worthy of publication, even when it has been obtained as part of a much wider programme of investigation. As Hardy characteristically put it: 'I doubt whether physics lends itself very well to treatment in terse and entertaining notes', his meaning being that pure mathematics *did*.[5] Sometimes mathematical writing was too compressed or too rooted in the local practice with which the author was familiar, because in contrasting mathematical cultures, such as those of Britain and Germany, reasoning that seemed clear to one mathematician might have been quite opaque to another. At the end of the nineteenth century this caused problems with the translation into German of Andrew Russell Forsyth's *Treatise on Differential Equations*, when the translator, unschooled in the Cambridge approach to mathematics, frequently had to ask Forsyth for an explanation of the examples.[6]

Consequently, there was no need for a mathematician to wait until he could present all his results as a connected and harmonious whole, rather than piecemeal. Indeed, the imperative worked the other way: the sooner a new result was published, the more certain was any claim to priority. Accordingly, such books as there were tended to be either textbooks for students, or treatises, which claimed to sum up the current state of knowledge concerning some branch of mathematics. A treatise could, of course, be an outstanding and scholarly work; Forsyth's *Treatise*, for example, was regarded as a masterpiece of synthesis, but it was not grounded in his original research. Even of treatises there was a distinct shortage: in 1886 the Cambridge mathematician James Whitbread Lee Glaisher complained that 'I could mention several subjects that are almost at a standstill, because advance is impracticable for want of avenues by which new workers can approach them', and added that the proportion of advanced mathematical treatises to memoirs was infinitesimal.[7] It was a different story with textbooks, for substantial demand both at home and in the Empire made them very profitable, and one of Isaac Todhunter's elementary works sold more than 600,000 copies.[8] However, there was no such incentive for the publication of advanced treatises, and so authors were frequently unable to earn what would have been useful additional income. Forsyth was therefore duly grateful to Macmillan & Co, who commissioned his first book, and when writing to them to celebrate 50 years of authorship he said that 'the co-operation of your firm is itself a contribution to research and becomes a crown to my labour'.[9] Nevertheless, it was not with books that advances were usually announced.

Returning to the *Jahrbuch*, the data suggest that through the second half of the nineteenth century there was a steady but undramatic increase in the number of British authors who published in pure mathematics, which resulted in an approximate doubling of their number over the period. The proportion of these who were members of the LMS fluctuated between 60 per cent and 80 per cent, with an average of about 70 per cent. However, in terms of output the contribution of members was greater than these figures would suggest, because from the 1870s onwards they wrote between 80 per cent and 90 per cent of the published pages. In broad terms, therefore, we can say that from soon after the foundation of the LMS those members who were authors in pure mathematics comprised a dominant authorial pool, with only about 15 per cent of the pages published by British authors being produced by non-members.

Data from the *Jahrbuch*, along with biographical sources, also yield information concerning the education of authors in pure mathematics. In this respect, despite the advent of the Universities of Durham and London and, as the century progressed, many other universities and colleges, the dominance of Cambridge increased: authors who had had their principal mathematical education there were, to an ever-greater extent, more numerous and more productive than those from all the other institutions combined. The insignificance of the University of London's degree in mathematics is remarkable, for it was offered not only in London but also in many places elsewhere in England and Wales. However, for the ambitious mathematics student the only achievement that really mattered was to do well in the highly competitive Mathematical Tripos at Cambridge; but as schools rarely taught the subject to the level at which studies at Cambridge began, it was a common pattern for students to take the London degree as a preliminary course, and then progress to Cambridge.[b] Indeed, for long periods the University of London acted purely as an examining body that imposed no requirement to have studied anywhere in particular, so it was not unknown for a student already at Cambridge to take the London degree as a practice run for the Tripos; one such was Richard Pendlebury, the senior wrangler of 1870, who took his London degree in 1869. It follows that a college of the University of London rarely provided the formative education of an author in pure

---

[b] For our purposes, such individuals have been treated as Cambridge graduates rather than London, because it was Cambridge that determined their mathematical interests and working methods.

mathematics. Furthermore, the same is true of the many other educational institutions where students could take the London degree, some of which became universities, or constituents thereof, before the turn of the century. Consequently, the combination of a Cambridge education and membership of the LMS came to dominate the Victorian pure mathematical scene. The data also confirm the natural supposition that academics formed the occupational group that produced the greatest output, with schoolteachers a long way behind in second place.[c] The dominance of academics increased as the century progressed, at the expense not only of schoolteachers but also of churchmen and lawyers; such a change, evident in many learned and scientific disciplines, reflected the decline in non-specialist and amateur involvement.

Finally, the public's perception of the LMS did not derive just from the reputation of its members and its social organisation, because the *Proceedings of the London Mathematical Society* was recognised internationally as Britain's leading mathematical journal. It was sent not only to members as a benefit of their subscriptions, but also to universities, libraries and other institutions at home and abroad, sometimes by way of an exchange of journals; and because it emanated from London, it had the advantage of presenting the Society to the world-at-large as being uncontaminated by the interminable internecine academic disputes at Cambridge. However, other outlets for pure mathematicians had been available before the formation of the Society. In particular, the *Cambridge Mathematical Journal* issued its first number in 1837, later added Dublin to its name, and in 1855 was finally reinvented as the *Quarterly Journal of Pure and Applied Mathematics*; there was also a minor journal, the *Messenger of Mathematics*. So with the Cambridge Philosophical Society and the Royal Society providing occasional forums for reading mathematical papers, pure mathematicians had not been unduly inconvenienced by the absence of a mathematical society. This was also largely true on the Continent, where other factors, such as the nature of more general scientific societies, were relevant. Consequently, although Continental mathematicians had embraced research since the eighteenth century, the Continent had very few mathematical societies.[10]

Once successfully established, the LMS was able to serve as a model in treating pure mathematics as a subject of high intellectual

---

[c] We define an 'academic' to be the holder of a chair, lectureship, fellowship or research position at a university or other institute of higher education.

endeavour that could benefit from an independent learned society, and it led directly to the foundation of the Société Mathématique de France in 1872. Similarly, the New York Mathematical Society, later to become the American Mathematical Society, was founded in 1888 by Thomas Fiske, who as a student at Cambridge had been befriended and introduced as a guest to the LMS by Glaisher, who was a member;[11] Fiske himself joined the LMS in 1893. In Germany the Deutsche Mathematiker-Vereinigung was not founded until 1891, largely at the instigation of the renowned and influential mathematician Felix Klein,[12] who was familiar with Cambridge and the British mathematical scene.[13] The late birth of such societies was not just a British phenomenon.

It is therefore unsurprising that forming a mathematical society was not an enterprise that naturally suggested itself to British pure mathematicians, because there was no flourishing Continental or American model from which the potential advantages could be assessed. This probably accounts for the unusually mundane origin of the LMS, which lay in a chance conversation in the summer of 1864 between two young friends, Arthur Cowper Ranyard and George Campbell De Morgan. Ranyard had just completed his mathematical studies at University College, London under George's father Augustus, and was about to embark on further study at Cambridge. However, once there the Mathematical Tripos examination showed him to be a mathematician of comparatively modest abilities, because in 1868 he was bracketed as twentieth in the list of senior optimes, or second-class-honours men. He was called to the bar in 1871, but his abiding interest throughout his life was astronomy, and he had been elected Fellow of the Royal Astronomical Society in 1863, when he was 18.[14] George De Morgan, who was a few years older, had earlier graduated under his father with high mathematical honours, but because of his delicate health did not proceed to Cambridge.[15] He became a mathematics master at University College School and Vice-Principal of University Hall, but died in 1867 at the age of 26.[d]

George De Morgan and Ranyard agreed that, in George's words, 'it would be very nice to have a Society to which all discoveries in mathematics could be brought, and where things could be discussed, as at the

---

[d] University Hall, which dates from 1848, was built in Gordon Square as a hall of residence for students of University College. It also housed Manchester New College, which had been founded in Manchester in 1786 as a non-conformist academy, and moved to Gordon Square in 1853.

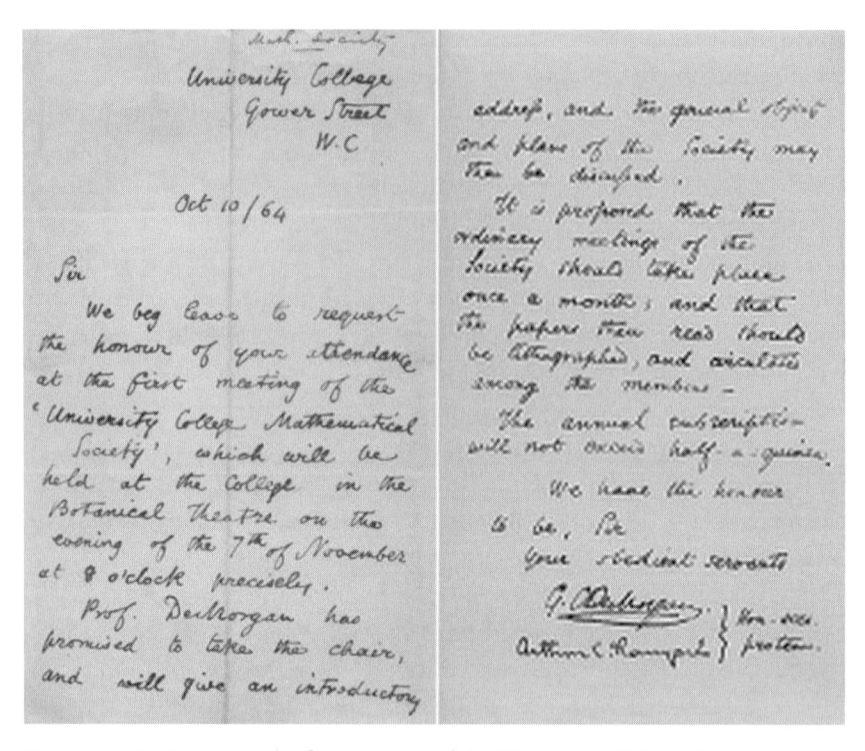

Figure 3.1 Invitation to the first meeting of the 'University College Mathematical Society', sent by George De Morgan and Arthur Ranyard to Thomas Archer Hirst.

Astronomical'.[16] That evening the proposal was discussed with Augustus,[17] who at Ranyard's request agreed to preside at the first meeting;[e] and so on 10 October the two friends sent an invitation to those connected with University College who might be interested. One recipient was Thomas Archer Hirst, who had been a mathematical master at University College School and was later to play a prominent part in the Society's activities. His invitation has survived; it announced 'the first meeting of the "University College Mathematical Society"', to be held on 7 November 1864 at the College's Botanical Theatre, and at which 'Prof. De Morgan has promised to take the chair, and will give an introductory address' (see Figure 3.1). It was further proposed that ordinary meetings should be held once a month, 'that the papers there read should be

[e] Letter from George De Morgan to Ranyard, 30 October, 1864, quoted in *Proceedings of the London Mathematical Society*, XXVI (1894), 554. The *Proceedings* refers to a bound volume of letters from George De Morgan to Ranyard concerning the founding of the Society, which Ranyard bequeathed to the Society in his will, but unfortunately there is now no trace of it.

lithographed, and circulated among the members', and that 'the annual subscription should not exceed half a guinea'.[18]

Given these events, it is natural to suppose that the Society was founded in 1864. However, the circumstances surrounding this first meeting have always been somewhat mysterious, and there are no formal records. In 1900 Robert Tucker, who had been a member since October 1865, on the Council since 1866 and a secretary since 1867 – and who therefore must have had many opportunities to discover what occurred – referred to 'the obscurity which now hangs over the early days of the Society'. He was not even able to state with certainty that the November meeting took place, saying only: 'We believe that a preliminary meeting was held ... '.[19] Fortunately, Hirst – who died in 1892 – kept a private journal to which Tucker presumably did not have access, and in which we find the following entry for 13 November 1864: 'On Monday last I attended the first meeting of the Mathematical Society at University College. De Morgan gave an address, which I seconded. I was put upon the Committee. I had at first declined but at De Morgan's request allowed my name to stand.'[20] This puts the fact of the meeting beyond doubt.

The name originally mooted by Ranyard and George De Morgan had been 'The London University Mathematical Society', but at the suggestion of Augustus it was amended to the 'University College Mathematical Society' before the invitations were sent.[21] However, this gave rise to adverse comments that 'the Society was only an upper higher senior class of De Morgan's',[22] which is probably what inspired a second change of name. Exactly when this occurred is unknown, although it was most likely to have been at the November meeting.[23] Perhaps these comments were also what led Augustus to do away with what Hirst described as 'the original restriction of membership to persons associated with University College',[24] notwithstanding that Augustus himself had made the proposal to include University College in the name. Whoever was responsible, by the date of its next meeting the Society had at last been named 'the London Mathematical Society'.

The air of mystery that surrounds the November meeting seems to have been almost deliberately cultivated, for Ranyard's obituarist in the *Proceedings* remarks that 'Ranyard was present, but all attempts on our part have failed to elicit a report of what took place.'[25] The situation has not been helped by a much-quoted article that Sir Edward

Collingwood wrote for the Society's centenary, in which he stated that 'Professor De Morgan happened to be ill and unable to attend the meeting . . . His place was taken by Dr. Hirst . . . '.[26] There was certainly a letter to Ranyard from 'De Morgan', quoted by the obituarist, that included the comment: 'You must make my excuses to every one for not being present',[27] but Collingwood appears to have overlooked that the letter was taken from a volume of correspondence between *George* De Morgan and Ranyard. The entry from Hirst's journal, quoted above, also supports this conclusion, since elsewhere he is punctilious in using George's initials or name to distinguish him from his father, even when no confusion could arise. However, although Hirst referred to an address given by 'De Morgan' at the November meeting, this was nowhere reproduced and was never mentioned again.

The underlying reason for the Society's coyness concerning its origins says much about the role that it saw for itself, as the following episode illustrates. Shortly after Augustus De Morgan's death in 1871 there was a move by his friends and pupils to perpetuate his memory with 'a bust and portrait, and . . . a handsome gold medal to be awarded by the Mathematical Society'.[28] Funds were raised, and eventually the organising committee decided that a triennial award should be made to some suitably worthy mathematician. In 1880 the Society formally accepted responsibility for making the awards and raising more funds to cast the medals, whereupon the die, which had already been commissioned, was handed over to the Society together with the remaining cash.[29]

It was during the second fund-raising operation that the Society circulated a letter in which Augustus De Morgan was described as the founder. This irritated Arthur Ranyard, who felt that the roles of himself and his deceased friend George De Morgan had not been properly acknowledged, so he complained to one of the Society's secretaries. The Council responded somewhat defensively by claiming that Augustus De Morgan, due to his early involvement, 'might be considered a joint founder'.[30] Ranyard thought that he had made his point, for he replied generously that 'there can be no doubt that the real founders of the Society were Prof. De Morgan and the other mathematicians of standing who early gave the Society their countenance. Without them it would have remained a students' society, and would probably have died a natural death.' This reply was quoted in Ranyard's obituary in the *Proceedings*, and a few sentences later the obituarist

(Robert Tucker) says: 'We have indicated what part he took in the inception of the *students'* society.'[31] The stress is in the original, and by emphasising this word and using the definite article Tucker revealed the Society's true motive for being so reticent about its origins. If that November meeting had been accepted as the Society's first, then undoubtedly it was founded in 1864 as a students' college society, a perception that was not at all to its advantage, for by 1871 it was even exploring the possibility of a royal charter.[32] However, Tucker also surmised that at the meeting 'Dr. Hirst ... proposed that the scope of the Society's operations should be enlarged', so if it was the one and only meeting of a students' society out of which emerged a plan for something more ambitious, and Augustus De Morgan was present, then it could be maintained that *he* was the founder of that more ambitious society. Furthermore, although the Society continued to be advertised in the Calendar of University College for two years after its foundation,[33] the membership, the standard of papers presented and the manner in which it conducted itself ensured that from the very beginning it in no way resembled a students' society. Everything was in keeping with a learned society founded by the highly esteemed Augustus De Morgan – provided that the true origin remained in obscurity.

Notwithstanding this, the unauthorised (but perhaps more correct) version was still circulating in 1892, when Hirst's obituarist in *The Times* referred to 'the London Mathematical Society, which was founded in 1864 under the Presidentship of Professor De Morgan, by students of University College'.[34] The compiler of *Alumni Cantabrigienses* also seems to have understood the situation, because he described Augustus as having been 'first President of the Mathematical Society, 1864'.[35] Nevertheless, Augustus accepted that 1865 was the year of foundation, as we can see from a badge that he designed for the Society, which includes this date. The badge also tells us something about the role that he thought the Society could play in the world, but although he must have spent some time pondering the design, he did not discuss it with his wife Sophia; so when after his death she found a sketch, she made a tracing and sent it to Hirst with an accompanying letter, saying that 'My son has done the little sage in the centre, who I suppose represents either Greek mathematicians in general, Aristotle or Euclid. The Society will understand the device, I have no doubt; but I cannot

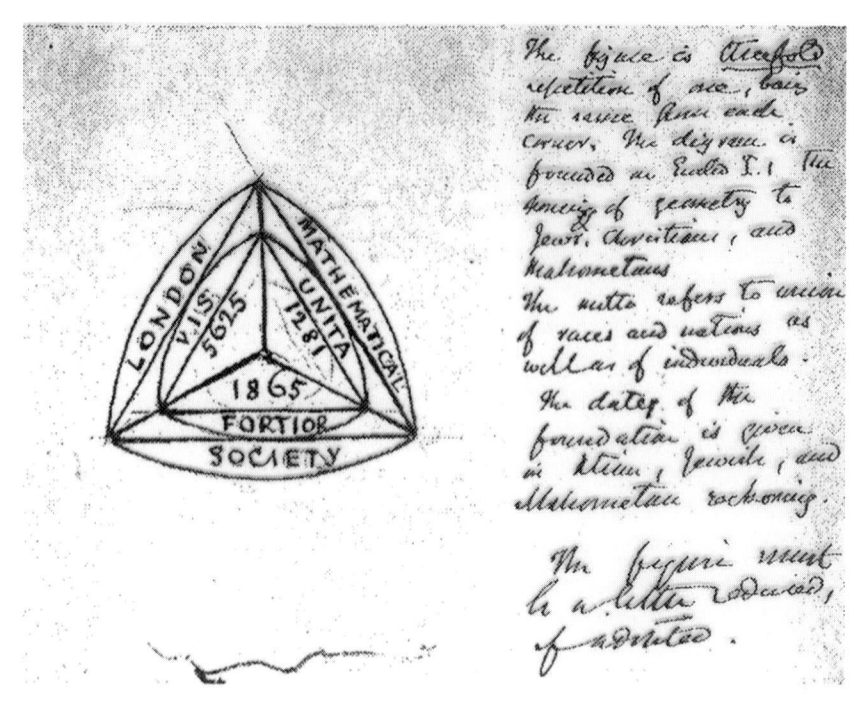

Figure 3.2 The original design by Augustus De Morgan of a badge for the London Mathematical Society, with his explanation.

quite make out the triangles & curves, which have a look of circle-squaring – nor the two dates at the sides, 5625 & 1280.'[36] Clearly she had never received any explanation from Augustus, which is why the design mystified her as much as it would probably have mystified everyone else, and there is no indication that the Society ever considered adopting it.

However, this was not the first version of the badge. After she wrote to Hirst she found Augustus's preliminary sketch and explanation, and sent it to Arthur Ranyard along with other miscellaneous papers (see Figure 3.2).[37] The explanation, in De Morgan's handwriting, reads:

> The figure is <u>threefold</u> repetition of one, being the same from each corner. The digram is founded on Euclid I.i the opening of geometry to Jews, Christians, and Mahometans The motto refers to union of races and nations as well as of individuals. The date of the foundation is given in Xtian, Jewish and Mahometan reckoning. The figure must be a little reduced, if adopted.

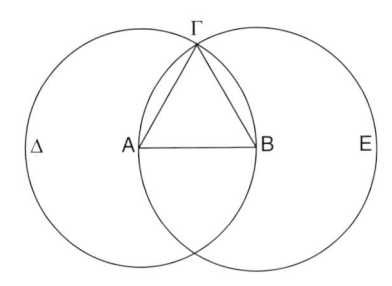

Figure 3.3 The construction diagram from the first proposition in Euclid's *Elements*, showing the origin of the badge designed by Augustus De Morgan for the London Mathematical Society.

This shows conclusively that De Morgan took the date of foundation to be 1865. He was sincere in his religious beliefs,[38] but the badge shows also the strength of his conviction that the LMS should disseminate mathematical knowledge amongst all people, without regard to religion, race or nationality: the motto '*Vis Unita Fortior*' ('A United Force is Stronger'), alongside the year in the calendars of the three major monotheistic faiths, is a declaration that the Society should use mathematics to break down boundaries between diverse peoples.[f] In particular, Euclid's *Elements* was assumed to be universally understood and appreciated, and De Morgan's reference to the First Proposition in Book 1, which shows how to erect an equilateral triangle with a given straight line as its base, also explains the graceless design of the badge, for it is taken directly from the proposition's diagram (see Figure 3.3). In De Morgan's sketch it is even possible to make out his faint construction lines.

So, rightly or wrongly, the Society has always taken the meeting on 16 January 1865 to be its true inauguration, and the 27 individuals who attended to be its founding members; all but one of them had some connection with University College or University College School.[39] Apart from Augustus, there were only a few who were, or were to become, distinguished in fields relevant to the Society: the names that stand out today are those of Hirst, a specialist geometer, but more important as a proselytiser; Edward John Routh, perhaps the most successful of all the Cambridge private tutors; and Robert Bellamy Clifton, who was to be professor of experimental philosophy at Oxford for more than 50 years.

---

[f] This Latin motto has now been adopted by the British Association of Urological Surgeons.

At the January meeting, Augustus gave an address in which he said that 'Our great aim is the cultivation of pure Mathematics and their most immediate applications. If we look at what takes place around us, we shall find that we have no mathematical society to look to as our guide.'[40] This language is somewhat misleading, because by 'pure Mathematics' he meant fundamental principles, and when he talked of 'applications' he did not mean 'applied mathematics' as we understand it today. He continued: 'In some quarters the mathematics are looked at, I may say almost entirely, with reference to their applications. These applications are not only physical applications or commercial applications, which may be termed *external*; but there are also what are rather *internal* applications.' He then gave the study of curves of the second order as an example of an internal application, and said that they 'are applications of the first principles of pure Mathematics.[g] They form, in fact, a particular branch of the applications of first principles. We must not mistake or misapprehend these internal applications – we must not regard them as constituting entirely what we are to turn our attention to.'

As this suggests, he was very exercised over the prevalence of 'internal applications', and the narrowness of view that could result: 'In the existing mathematical journals and publications, the offered inducements are, for the most part, confined to one particular class of subjects.' Such a journal was the *Quarterly Journal of [Pure and Applied] Mathematics*, by which 'many are repelled' because of its specialisation. Even worse was a journal that he referred to as the 'Woolwich Journal of Mathematics'; almost certainly this was *The Mathematician*, which had had a brief career from 1843 to 1850.[41] It was edited by the staff of the Royal Military Academy, Woolwich, who also provided most of the content, and De Morgan complained that it had been given over in great measure to the properties of conic sections: 'before you could find a bit connected with the Differential or Integral Calculus, you would come across half a dozen applications or proofs of Pascal's theorem'.[h] It was to this fondness for conic sections that De Morgan attributed *The Mathematician*'s early demise.

---

[g] 'Curves of the second order' are the conic sections (circle, ellipse, parabola and hyperbola), together with the equations that represent them in coordinate geometry.
[h] Pascal's theorem treats of the lines joining six points taken at random on a conic section.

So when he said of the Society that 'Our great aim is the cultivation of pure Mathematics and their most immediate applications', he was not only excluding 'external applications', but also giving priority to the cultivation of 'pure Mathematics' – that is, the first principles of mathematics; and this would be enhanced, he believed, by paying attention to four special topics of particular interest to students: logical mathematics, which he defined as the connection between logic and mathematics; the history of mathematics; language; and modes of proof. Although he stressed that 'people must be pleased', he did not anticipate yet another club dedicated to generating and solving problems; rather, he expected that the Society would be of a novel kind, catering for mathematicians who wished to support pure mathematical research as an independent discipline.

Notwithstanding this, he thought that the LMS should have a broad appeal, and must have '*the great bulk of its business adapted to the great bulk of its members*. In order to do this, those versed in the higher Mathematics, in Canonical Algebra, Elliptic Transcendentals, and a dozen other such subjects which have been opened lately, must study the inclinations and pursuits of those engaged on the more elementary branches.' This would require the most mathematically advanced members to act altruistically, not only by discussing subjects outside their own particular specialities, but also by considering the needs of those less able, or less forward in their studies, than themselves. The LMS would thereby be distinguished both from the Royal Society, which 'may be said to contain a mathematical section' but was not a mathematical society, and from the Cambridge Philosophical Society, which was something nearer what he had in mind, but was (he said rather pointedly) 'in an exceptional position ... it is a society in which almost all the members are able to relish its highest discussions'. He was suggesting here that the LMS, rather than catering exclusively for such a clientele, should take a more relaxed approach, and he emphasised the importance of personal interaction for the pursuit of mathematics. However, that also implied something about the aims and objectives of the Society, which should encourage the participation of a wide range of members, possibly at the expense of its highest intellectual ideals. We shall see later the extent to which his vision was realised, but within a year he considered that what he had set out as his cardinal principle – that 'people must be pleased' – was in danger, for he issued a reminder that 'papers should be kept within the comprehension of the majority of the hearers'.[42]

Although the Society existed fairly informally throughout 1865, it nevertheless recruited three of its most mathematically distinguished members: Arthur Cayley, the Sadleirian Professor of Pure Mathematics at Cambridge, Henry John Stephen Smith, the Savilian Professor of Geometry at Oxford, and James Joseph Sylvester, always something of a loose cannon but at that time professor of mathematics at the Royal Military Academy, Woolwich. Cayley's importance to the Society as the most highly regarded pure mathematician in Britain can be judged from the fact that in 1868 it abandoned a fixed date for its monthly meetings in favour of giving the Council the right to set whatever date it wished; this, says Glaisher, was so that Cayley could attend the meetings of the LMS and the Royal Astronomical Society on the same visit to London.[43] However, it was Sylvester who provided the *Proceedings* with its first mathematical paper, which turned out to be of considerable interest, for he was able to prove and generalise a rule of Isaac Newton's for the discovery of certain roots of an algebraical equation.[44]

For the first year of the Society's existence the monthly meetings typically attracted 15 members, a comparatively high figure that was to decline during the succeeding decades. Notwithstanding this, during 1866 the membership grew to more than 100, which required a reconsideration of the Society's mode of operating. It was still meeting at University College, but this was not satisfactory and so an agreement was reached with the Chemical Society for the use of its rooms at Burlington House from November 1866.[45] Unfortunately the new accommodation still did not give the Society space for the small library that it was accumulating – its nucleus had been a gift of books from Sir John Lubbock in October 1866[46] – which continued to be housed at University College on the shelves of William Sharpey, the professor of physiology.[47] In 1870 the rebuilding of Burlington House forced a further move to premises leased from the Asiatic Society in Albemarle Street, which gave the Society two rooms for council and general meetings, and a further small room for the library. Although the rooms were 'taken from year to year'[48] the Society was to remain there for the rest of the century, and beyond.

The move to Burlington House, and Sylvester's taking the presidency in succession to Augustus De Morgan, began to loosen the informal but close association between the Society and University College. In 1865 this association had troubled Samuel Roberts, one of the Society's earliest recruits, and it was still troubling him 26 years later when he wrote: 'At the outset I raised a feeble voice against continuing at U. C. I still think every stone should be turned to avoid being associated in the mind of the Mathl. World with any educational institution in particular.'[49] As we have seen, this was the reason for the second change of name during the Society's formation. Even so, it is likely that for some members such as Roberts 'the London Mathematical Society' was still too easily taken as referring to the university. Consequently the Society was often, and semi-officially, referred to differently. Sylvester's paper on Newton's rule was described in the *Proceedings* itself as being 'for the use of the Mathematical Society of London', which was unambiguously a geographical rather than an institutional reference, and pleasingly concordant with 'the Royal Society of London' – perhaps looking forward to the day, not so far ahead, when the LMS would feel sufficiently confident of its status to seek a royal charter. 'The Mathematical Society' was also frequently used as an appellation, and only in 1894 did the Council decide that 'the words "The London" should be painted above the words "Mathematical Society" on the Society's entrance door'.[50]

The Council meeting in March 1866 passed a resolution that had been proposed by Hirst: 'that all papers read before the Society with a view to publication be referred to two members of the Society to be approved by the Council, who shall report thereon to the Council'.[51] This was the beginning of the Society's commitment to strict refereeing, which was modelled on the procedure at the Royal Society and thereafter rigidly applied:[52] if the referees were in broad agreement then a ballot of the Council decided the question of publication, but if the referees disagreed then the paper was sent to a third referee for arbitration.[53] Although the Council made the final decision, it rarely sought to override the judgement of the referees, for the Society was always determined that the papers published in the *Proceedings* should be prepared and assessed according to the highest academic standards.

Furthermore, the LMS had no truck with fringe mathematicians such as circle-squarers, who attempted a feat that was widely considered impossible, namely the geometrical construction – using only straight-edge and compass – of a square having the same area as a circle;[i] so when in 1867 it was offered 'an infallible method of squaring the circle', the secretary was 'requested to inform the writer of this letter that the

---

[i] The impossibility of squaring a circle was proved in 1882 by Ferdinand von Lindemann.

Society did not entertain such subjects'.[54] The Council was here following common practice: in 1861 James Smith, a dedicated circle-squarer, noted that 'The [British] Association, following in the footsteps of other learned bodies, declines to permit any discussion on this subject to be introduced into its deliberations,'[55] and in November 1871 the Society formally resolved not to accept papers dealing with any of the three great unsolved problems of classical geometry, namely squaring a circle, constructing a cube having twice the volume of a given cube, and trisecting an arbitrary angle.[56] Years later, however, this did not deter an Indian gentleman from offering to forward a complete solution of the trisection problem upon receipt of £100,000.[57]

At the Council meeting in April 1866 a further initiative by Hirst was to propose: 'That persons being neither British subjects nor residing in Her Majesty's dominions shall be selected from among mathematicians of the greatest eminence for Honorary membership';[58] thus began the Society's attempt to make international connections. It is perhaps surprising that Augustus De Morgan, in his opening address, made no reference to the relevance that such connections might have for the Society; instead, he was almost entirely concerned with its internal nature and the conditions necessary for its survival, so he treated not at all of its relationship with the rest of the world. This new class of membership was supposed to bring lustre to the Society, which nevertheless made no effort to appoint any such members until the following March, when Sylvester reported that the distinguished French geometer Michel Chasles wished to be elected an ordinary member. The Council, seizing its opportunity, at once resolved that he be nominated a foreign member – a term that the Society used synonymously for honorary member[j] – and the secretaries were 'instructed to prepare a form of diploma, taking that of the Royal Society as a model'.[59] Sophia De Morgan told how her mortally ill son George, who 'was still warmly interested in the success of the Mathematical Society', should have signed the diploma as one of the secretaries: 'For this purpose parchment was placed before him, and he evidently recognised its import, passing his fingers over the words Mathematical Society. But he was too weak to hold the pen, and died two days after.'[60]

This appointment turned out to achieve more than the Council anticipated, although not for the Society. Chasles, a geometer of great

---

[j] To avoid confusion, they will henceforward be referred to as honorary foreign members.

European distinction, was at that time under considerable pressure in the French Academy because he had championed the authenticity of a large number of purportedly old but very dubious manuscripts that he had purchased. He became the target of violent criticism, and eventually his supplier, Vrain-Denis Lucas, was convicted of forgery. News of the *affaire* reached England in the autumn of 1867, but had it been a few months earlier it is questionable whether the Council would have been so eager to grant Chasles his honorary foreign membership, particularly as some of the manuscripts appeared to prove that it was Blaise Pascal, rather than Isaac Newton, who first discovered the laws of gravitation. There were also transcriptions of letters purportedly written by Cleopatra and Mary Magdalene, amongst others, and both at the time and in retrospect Chasles appears to have been extraordinarily gullible.[k] Fortunately his mathematical reputation survived intact, and as a result he became an ardent admirer of the LMS, in which a person's standing was judged solely by mathematical achievements – very different from his experience in the French Academy. However, there was no such society in France, so when he came to write an official report on the progress of geometry he concluded by making a plea for the establishment of a French national mathematical society, to prevent Britain from gaining an advantage.[61] His suggestion bore fruit, and the Société Mathématique de France was founded in 1872 as a direct consequence of his appeal.[62]

After the first 18 months the increase in membership meant that the LMS needed a more formal set of rules than that which had obtained hitherto. A rulebook was drawn up by Hirst and Sylvester for presentation at a Special Meeting of the Society on 15 October 1866, and it was 'passed with certain alterations';[63] these rules underpinned the Society's operations for the rest of the century. Rule 36 in particular is both interesting and confusing, because it attempted to express the Society's attitude towards 'applied Mathematics', whatever that term was taken

---

[k] There is disagreement concerning the name of the forger. In 1902 Charles Whibley (*The Cornhill Magazine*, 85, p. 628) referred to 'Vrain-Denis Lucas – or, as he was commonly known, Vrain Lucas...', and contemporary French references were to Vrain Lucas or Vrain-Lucas. Reports in *The Times* mention Lucas Vrin (18 February 1870, p. 9) and Vrin Lucas (19 February 1870, p. 8). The present author follows Whibley and Rebekah Higgitt ('"*Newton dépossédé!*" The British Response to the Pascal Forgeries of 1867', *The British Journal for the History of Science*, 36, 4 (2003), 437–453) in referring to Vrain-Denis Lucas, but many modern references are to Denis Vrain-Lucas. Higgitt's paper gives a detailed account of the English reaction to the claims made by Chasles.

to mean; the rule provided that 'At no two successive Meetings shall the papers be entirely on applied Mathematics.' In 1874, when the Society was attempting to attract new members, it was noted that that the rule concerning applied mathematics allowed up to one-third of evenings to be devoted thereto, and that 'Hitherto nothing like this proportion of time has been so occupied. It might perhaps conduce to the interests of the Society if the applications of mathematics occupied a somewhat more prominent place than they do at present.'[64] It is not clear how the reference to 'one-third' was derived, because at face value the rule allowed almost any amount of time to be spent on applied mathematics. Nevertheless, however the rule was interpreted, the Council clearly thought that the Society already had too much of a reputation as a forum primarily for pure mathematicians, and that it could make itself more attractive to potential members if it extended an equal welcome to applied mathematicians. Rule 36 was accordingly abolished in December 1874,[65] and thereafter there was no formal restriction on the number or nature of papers that could be presented.

It seems that the only way of making sense of the rule is to assume that the term 'applied Mathematics' was originally used to implement De Morgan's warning against being overwhelmed by 'internal applications' such as curves of the second order, but that later it was being interpreted in its modern sense of his 'external applications', which suggests considerable uncertainty over terminology. However, this does not explain the reference to 'one-third of evenings', which remains mystifying. It has also been suggested that the rule was abolished out of respect for one of the members, the mathematical physicist Lord Rayleigh.[66] Therefore it is likely that in abolishing the rule the Society was not only (it hoped) becoming more welcoming to potential members, and perhaps attending to Lord Rayleigh's feelings, but also recognising that language had changed, and that the use of 'applied Mathematics' with De Morgan's meanings was by then outdated and confusing.

Although there was a clear need to boost membership, this competed with a desire to maintain the highest standards, so in spite of De Morgan's hopes the Society made little effort to attract students or new graduates. Many years later James Whitbread Lee Glaisher, who was eventually a member for more than 50 years, recalled that 'I took my degree in 1871, and was elected a Fellow of the Royal Astronomical Society in April of that year; but I was too modest to ask to be proposed as a member of the London Mathematical Society, which seemed to consist almost wholly of men whose mathematical or scientific position was already established.'[67] This preponderance of older mathematicians was not surprising, as Glaisher goes on to say that at Cambridge 'no attempt had been made to reach the younger lecturers and others.'[68] Indeed, to try to drum up students or new graduates as members would have been considered unseemly for such an august body as the LMS.

Nevertheless, the Society did continue to grow. Although membership remained roughly flat at between 100 and 119 for the first 9 years, it then increased steadily at an annual rate of about 3 per cent, which compares well with the early history of some other societies such as the Astronomical, the Statistical, and the Institute of Actuaries. However, societies dealing with the natural world, or instituted for utilitarian purposes, were usually more successful than those in the mathematical or physical sciences; for example, the Institution of Mechanical Engineers was founded in 1847, and after bumping along for nearly a decade with a membership of around 210 began a period of straight-line growth to about 1,200 by 1880, which represents a year-on-year growth-rate nearly double that of the LMS.[69] The significance of the Society's expansion lies in the small numbers involved: considering that Wallis and Wallis identified more than 19,000 individuals working at some time during the eighteenth century who could be described as 'mathematicians', it is remarkable that by the end of the nineteenth century ordinary membership of the LMS had reached only 249.[70] This highlights how terms such as 'mathematics', 'mathematical' and 'mathematician' had become very much narrower in meaning.

Because the LMS began in London, that is where most of its earliest members lived. However, the membership rapidly acquired a broader geographical base, and 1867 saw the election of Martin Gardiner, a resident of Australia, who became the first overseas ordinary member. If we define a member-year as representing one member being a member for one year, then over the whole period until the turn of the century about one-tenth of the total member-years came from members living abroad. However, the only overseas countries with a significant number of members were India and the United States, and there was virtually no representation in Continental Europe, which is where the most interesting developments in pure mathematics were taking place: from 1865 to 1899 there were only five ordinary members with Continental addresses, and two of the five were British nationals.

In the United Kingdom, the principal concentrations of members were in Cambridge, Dublin, London and Oxford, and throughout the Society's first 35 years, during which membership of the Society more than doubled, the number of members having London addresses remained virtually the same. On the other hand, the number of members living in Cambridge increased more than fourfold to almost the number living in London, which gave it a very much greater concentration of members than any other city or town in Britain. Although this purely geographical analysis says nothing about the academic qualifications of the members, such an inference would not be misleading, for in spite of its name the LMS rapidly transformed from a society that originated with the students, graduates and staff of University College, London, into one dominated by Cambridge graduates. This was almost inevitable, for although Cambridge University had shortcomings as a source of new mathematics, the most able mathematics students naturally gravitated there due to its unrivalled reputation as a place to study, and so its graduates provided the bulk of the Society's members. In contrast, the number of members who received their principal mathematics degree from the University of London actually decreased, and for the rest of the century was never greater than in 1867. For the same period, Cambridge graduates contributed 10 times more member-years than did London graduates.

An early exodus of London graduates is not unexpected, because the original members naturally included some who joined more as a gesture of support, or out of curiosity, than through any particular commitment to a mathematical society. Such members soon drifted away, and even Arthur Ranyard lost interest, although he remained a member until his death: Glaisher reported of him that 'He was a very active member of the Council of the Astronomical Society, and he and I were colleagues for a time as joint Secretaries; but I never heard him mention our [Mathematical] Society.'[71] The Society's problem was its failure to attract new members from the increasing pool of London graduates who did *not* proceed to Cambridge, a pool swollen by those who had studied at provincial institutions that offered London degrees in mathematics. The expansion of university education in the second half of the nineteenth century had no impact on the increasing dominance of Cambridge graduates over the LMS: in 1867 they already constituted about half the total membership, but by 1897 the proportion had increased to almost two-thirds, and at the end of the

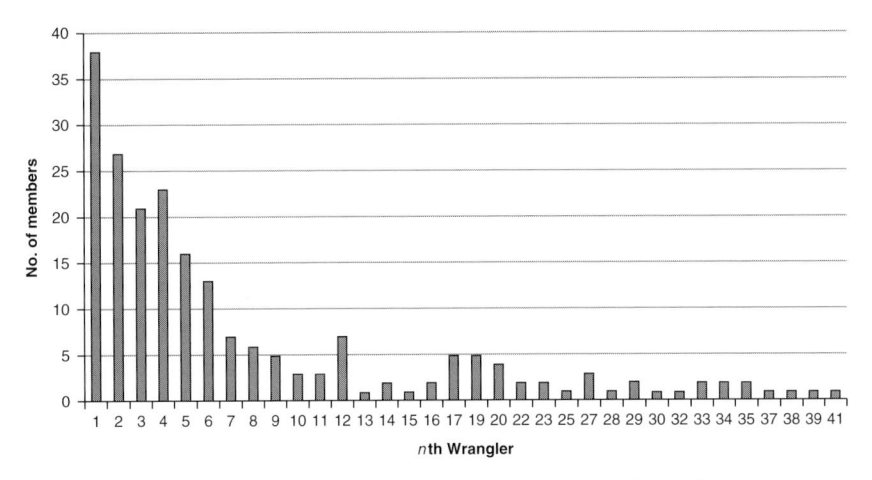

Figure 3.4 Chart showing the total number of ordinary members of the London Mathematical Society at each position in the table of wranglers during the period 1866–1899.

century, out of a total of 249 members there were only seven whose principal degree was from London.

We can investigate the predominance of Cambridge a little further, thanks to the public listing of Cambridge wranglers in order of merit. During the period up to 1899 there were 443 individuals who were at some time ordinary members, of whom 245 were Cambridge graduates. They included 212 wranglers, and the chart in Figure 3.4 shows that the Society was most attractive to the highest ranked in this group, with interest rapidly declining amongst those who did not achieve such honours. Viewed as potential members, there were about the same number of wranglers at each position in the order (except at the tail-end), yet 38 senior wranglers joined the Society, compared with only one fifteenth wrangler; and no eighteenth wrangler joined at all. In fact, of the 212 wranglers, only 74 were seventh wrangler or lower. Of senior and junior optimes there were very few.

There are two explanations. The first lies in the intellectual level at which the Society sought to operate, as shown by the content of its *Proceedings*. This was for most members their only contact with the Society, which almost from its inception had rules that aimed for the highest standards in the refereeing and acceptance of papers; the purpose, not particularly difficult and soon achieved, was to make the *Proceedings* the country's foremost mathematical journal. Augustus

De Morgan had suggested that the LMS should set its sights somewhat lower than the Cambridge Philosophical Society, but in practice it did not do so. Consequently the Society's strongest appeal was to those most familiar with mathematics at a very advanced level; that is, those who had achieved the greatest success in the Mathematical Tripos. The second explanation is that those who did not achieve such success were obliged to leave Cambridge and seek a full-time career elsewhere, after which new interests would naturally supersede the involvement with mathematics of all but the most dedicated. On the other hand, the top wranglers, most of whom continued at Cambridge for at least a few years, had more than their success in common, for many of them inhabited the same environment and would have been in regular contact with each other when undertaking their college duties or socialising. This was particularly so because the adjoining colleges of St John's and Trinity had an exceptional reputation for producing wranglers, and between them contributed about one-third of the Society's membership; these members had spent 10 terms, and often many additional years, in the same dozen acres.

The situation regarding Oxford graduates was very different, for in spite of having Henry Smith – possibly the British mathematician most in tune with Continental developments – as Savilian Professor of Geometry, the university had no mathematical reputation, and its overall contribution of 46 ordinary members was minor when compared to that of Cambridge, as were the contributions of London with 33, Trinity College Dublin with 24, Edinburgh with 2, Glasgow with 1, and Queen's College Galway with 1. Of the remaining ordinary members, 32 had degrees from overseas, 24 had no degree, and there were 35 obscure individuals for whom no information is available.[1] So although the University of London provided the principal degree of an increasingly large pool of potential members, the Society failed to take advantage of this, and the representation of the university in the Society declined substantially, both absolutely and proportionately, until by the turn of the century there was nothing to suggest that it had ever been significant. Notwithstanding this, when in 1886 the Society debated a proposal that the word 'London' should be omitted from its

---

[1] The existence of comprehensive data for British universities suggests strongly that these obscure individuals either had degrees from overseas institutions, or (most likely) had no degree at all.

name, it was almost unanimously defeated.[72] Nor did it ever consider relocating; it sought status and hankered after a royal charter, ambitions that would have been severely compromised by moving out of London, for it would thereby have declared itself to be only on a par with other provincial scientific and learned societies.

This elitism is evident from the *Proceedings*, to which the Society attached very great importance, and it is sometimes assumed that the journal supplied a need, because it was the first in Britain of such a high standard. However, a lack does not imply a need, and there were many avenues for publication open to a determined and talented mathematician. The outstanding example is Arthur Cayley, for he was the most productive of all British pure mathematicians, as well as being considered at the time, by common consent, to be the best of his British contemporaries; by the end of his life he had published more than 900 papers, in some of which he opened up new avenues for algebra that are now part of every mathematician's armoury. After his triumph in the Tripos and Smith's Prize examination of 1842, he remained at Cambridge for three years as a tutor, but then read for the bar to qualify as a barrister. The law was to be his profession for almost 20 years until in 1863 he was elected the first Sadleirian Professor of Mathematics at Cambridge.

Between 1842 and 1863 Cayley produced 308 papers, which appeared in 20 different journals. About one-fifth of this output was published overseas – mostly in *Crelle's Journal* and Liouville's Journal – with the remainder appearing in British journals of various kinds. In view of Cayley's reputation as a very pure mathematician indeed, it is worth remembering that he made significant contributions to the *Memoirs* and *Monthly Notices* of the Royal Astronomical Society, of which he was a keen member – he edited the *Monthly Notices* for more than 20 years. His great output shows that there was no insuperable difficulty in the way of a first-rate British mathematician finding a home for his papers; furthermore, he was practising law throughout nearly all of this period, and so was – on a strict construction of the word – an amateur for whom publication might have been more difficult. Nevertheless, although there were many possible outlets for British mathematicians, it was probably desirable for there to be a dedicated British journal that could compete with those on the Continent. By this means, the best of British mathematics would be given a natural home, and British mathematicians – of whom only Sylvester and Henry Smith could rival Cayley – would have an internationally recognised platform for their work, rather than having it scattered around in sundry journals that were not necessarily dedicated to mathematics.

The content of the *Proceedings* confirms that British mathematics increasingly meant Cambridge mathematics, because the majority of the pages that were written by ordinary members came from Cambridge graduates, rising from about one-half in the 1870s to three-quarters in the 1890s, peaking at over 90 per cent in 1899. Of course, this was partly the result of the increasing proportion of ordinary members who *were* Cambridge graduates – they comprised almost all the new intake during the 1890s – but was also because, as a group, they were more prolific than the other members. However, a threefold increase in the length of annual volumes was mainly due not to the number of papers, but rather to their average length, which more than doubled. This supports the complaint of E B Elliott, when president in 1898, that the length and complexity of the papers being read at meetings were adversely affecting the social function of the Society; he remembered Henry Perigal, the Society's most aged member who had recently died in his ninety-eighth year, and who:

> used at our meetings in the young days of the Society to help it in one of its aims – not so much, I regretfully realize, to the fore now as was the case then – of making our gatherings together occasions for informal interchange of ideas, and consequent stimulation of the comparatively unlearned among those who had the cause of mathematics at heart...

On the same theme, later in his address Elliott was concerned that:

> the advantages to be gained from [the Society's] meetings themselves by a large body of its members should diminish, that so many of us should never attend a meeting of the Society at all, that the reading, or taking as read, of memoirs of comparatively ambitious character presented for publication in our *Proceedings*, often from their nature and length difficult to state orally in lucid abstract, should come to be regarded as alone appropriate to the grave dignity of our gatherings.[73]

In referring to 'memoirs of comparatively ambitious character' Elliott doubtless had in mind the longest paper to have been published in the

Proceedings up to that time, which was by Alfred Greenhill in 1893 – it ran to 109 pages.[74] Augustus De Morgan believed that encouraging personal engagement between members should be an important aim for the Society, yet clearly it had fallen by the wayside. Furthermore, the Society failed to connect with the general public; the nearest it came was in 1876, when it loaned Plücker's models of curved surfaces for an Exhibition of Scientific Apparatus at the South Kensington Museum.[m] Even then, the original intention that Olaus Henrici, an ordinary member, should explain them in a lecture was never fulfilled.[75]

However vigorous the Society's social function may originally have been, Elliott was right in saying that it diminished as the century progressed. The turnout at meetings was always low, and usually included less than a quarter of those members who lived in London and so could have attended. This was partly due to the nature of mathematics, which even at the best of times is a difficult subject for oral presentation, and was getting ever more technical: Glaisher, when making just this point in 1926, recalled an occasion decades earlier when Samuel Roberts struggled to make himself understood, while the other members, who had quite lost the thread, immersed themselves in the journals that were lying around the meeting room.[76] Nor did the Society hold local meetings in the manner of the British Association, and so for most members the *Proceedings* provided the main benefit of membership. It was always a vehicle for British mathematicians, who until the turn of the century provided more than 95 per cent of the content, and it never achieved the status of Continental journals such as those of Crelle or Liouville. These attracted a much more international pool of authors because Continental mathematicians, to a far greater extent than the British, participated in a mathematical culture that straddled national boundaries.[77] In Britain the culture revolved around Cambridge, with one generation of Cambridge-trained mathematicians teaching the next, and often their work was not always easy to understand for those who had been educated elsewhere, due to the emphasis in the Tripos on solving short, concise problems, and to the tacit knowledge shared by Cambridge authors and readers. This was true even in respect of British universities, and when comparing London's

---

[m] These models are now on display at De Morgan House in London, the Society's headquarters.

University College with Cambridge, Andrew Warwick described 'the pedagogical incommensurability that existed between these two geographically separated places of mathematical training'.[78]

As regards what was published in the *Proceedings*, the clearest trends are a dramatic decline in the space given over to geometry, an increase in the popularity of algebra, and the consistently strong showing of applied mathematics, which attracted many papers; indeed, from 1895 to 1899 nearly one-third of the pages were given over to it. Glaisher's comment that 'It was a constantly recurring source of regret during the whole time that I was on the Council that the Society did not attract to itself papers on Applied Mathematics'[79] is therefore puzzling, as he was on the Council for 33 years from 1872. The inconsistency between his comment and the number of papers published is not because we now regard as applied mathematics topics that he considered to be pure, so perhaps his regret was that there were not *more* papers on applied mathematics. It is certainly true that pure mathematics dominated the *Proceedings*, but nevertheless applied mathematics was always in evidence and increased in popularity throughout the period. In the new century, however, there undoubtedly was a decline in the attractiveness of the *Proceedings* for applied mathematicians.

The Society gained additional exposure for the *Proceedings* by distributing free copies to authors and libraries, and also by engaging in an exchange of journals with other societies, provided that the proprieties were observed; so when the new 'Mathematical Society of Paris' was formed in 1872, the Council of the LMS decided 'that overtures for an interchange of Proceedings should come from the younger Society'.[80] Furthermore, the Society always tried to ensure that the *Proceedings* would be of real value to the recipient, but notwithstanding this caution by 1893 it was distributing to libraries and other societies 49 free copies of the *Proceedings*, with 34 of them being sent outside Britain.[81]

Such was the importance attached to the *Proceedings* that the Society, contrary to the intention of its student founders, soon came to believe that publishing, rather than holding meetings, was its main function, and this was the cause of financial problems that beset it almost from the beginning, for the cost of printing was greater than the subscription income. The founders have been accused of 'surprising naïveté' in setting the subscription so low,[82] but the original invitation, in which a maximum subscription of half-a-guinea was set, had not foreseen the use of high quality commercial printing, and proposed only lithographic reproduction; George De Morgan even suggested that the Beadle at University College should undertake this.[83] Furthermore, the proposal was for a students' society, so the subscription had to reflect the traditional impecuniousness of students, and was by no means modest when compared to those of some other societies at University College.[84]

Although at the general meeting on 15 January 1866 Robert Tucker, one of the secretaries, had reported that 'the finances were in a prosperous condition',[85] this was based on a receipts and payments account that showed a balance in hand of £21. 4s. 9d. but did not take account of outstanding printing costs.[86] By November the Society was insolvent, owing more to the printer than it had in the bank,[87] and it survived only by postponing payment. A year later the bank balance was just £1,[88] so the immediate step was taken of increasing the annual subscription from ten shillings to one guinea, in the hope of averting a winding-up.[89]

This measure temporarily staved off trouble, but the Society was now painfully aware of the need to generate additional income, and so decided to sell copies of the *Proceedings*. It was a successful venture,[90] but the costs were still uncontrolled and in December 1873 it was reported that expenditure for the year had exceeded income by £60, of which £50 was attributable to an escalating printer's bill. A sub-committee recommended that referees be asked to consider whether papers were fit for the *Transactions* (*sic*) and if so whether they could be shortened or abstracted for publication. A ban on purchasing books and journals for the library was also suggested.[91] However, before these measures could be assessed and implemented Lord Rayleigh made a gift of £1,000 to the Society, the income from which was to support the publication of the *Proceedings* and the purchase of journals.[92] This condition was hardly onerous because there was very little other expenditure, so the Society was immediately placed on a much sounder financial footing. In 1890 a gift of £500 from Lt Col Campbell was a further addition to the Society's investments,[93] and by the end of the century investment income comprised about one-third of the total receipts, with the balance coming equally from subscriptions, and sale of the *Proceedings*. However, although the Society always exercised great financial prudence, it was reluctant to press for arrears of subscriptions, a matter that caused considerable embarrassment, particularly as the Council was far too gentlemanly to take firm action. There seems no doubt that some members took unfair advantage, and this problem was never properly solved.

A desire to avoid anything that might entail unpleasantness or dissension was also apparent in the Society's response to a proposal that questioned the supremacy of Euclidean geometry, considered as a pedagogical tool. In June 1868 James Maurice Wilson, a schoolmaster who had only a few minutes earlier been admitted as a member, read a paper on 'Euclid as a Textbook of Elementary Geometry' – his argument being that it was highly deficient, particularly when compared to his own system, which replaced Euclid's basic ideas of 'point' and 'line' with that of 'direction'. The debate was nothing to do with non-Euclidean geometry, but centred on the merits of Euclid's *Elements* for teaching Euclidean geometry in schools and colleges. Wilson's contributions to geometrical teaching were perhaps of a more fundamental character than would normally have been expected from a schoolmaster because he had been senior wrangler in 1859. However, the normal academic progression to a lectureship or coaching position at Cambridge was denied him because of a total breakdown in his physical and mental health following the Mathematical Tripos, as a consequence of which he forgot virtually the entire corpus of higher mathematics.[94]

Wilson did not claim to be the first to offer a thorough criticism of Euclid, but awarded that palm to Augustus De Morgan, who, he said, contributed 'a subtle and original paper' to the *Companion to the British Almanack* for 1849.[95] The *Proceedings* records that 'Prof. Hirst, Messrs. De Morgan, Ellis, Hemming, and A. Smith took part in the discussion upon [Wilson's] paper, and Mr. Wilson replied',[96] but this is probably a diplomatic rendering of what occurred because the question was an emotive one, and more than half-a-century later Wilson still remembered being 'well "heckled"'.[97] The meeting may have had in mind the unhappy example of Italy, where 'the abolition of Euclid had led to such pernicious confusion in the teaching of various geometrical systems that the re-establishment of Euclid was decreed by authority'.[98] Wilson's paper was not published in the *Proceedings* but appeared instead in the *Educational Times* that September.[99]

A disillusioned Wilson concluded that the LMS was no place to air such controversial views, and, more importantly, that it was not interested in debating the practice of teaching. Accordingly, by 1871 he and his supporters had founded a new organisation. The name originally proposed was 'The Association for the Reform of Geometrical Teaching', but this was thought to suggest too specific an aim, and so 'The Association for the Improvement of Geometrical Teaching' (AIGT) was the name finally adopted.[100] By 1897 it had widened its scope considerably and was renamed 'The Mathematical Association'; it remains the foremost British society for mathematics teachers, and its existence is a consequence of the desire of the LMS to avoid dispute and confrontation at any cost. It was not that individual members of the LMS were necessarily hostile to the AIGT – Hirst served as its first president, and Sylvester had said that 'I should rejoice to see ... Euclid honourably shelved and buried "deeper than did ever plummet sound" out of the schoolboy's reach'[n] – but participation in such controversy was not seen as being helpful to the aims of the Society.[101] The result was that it distanced itself from matters of great relevance to many members and potential members, and Wilson himself resigned in 1874. De Morgan's dictum that 'people must be pleased' clearly had its limits.

This debate suggests the limited role that the Society thought it should play. The topic of mathematical education was one of general concern, and nine members of the LMS supported Wilson by immediately joining the AIGT; from these nine were drawn not only the president but also a vice-president, the treasurer, and five local secretaries, one of whom (Robert Tucker) was also one of the secretaries of the LMS. Yet of the 61 members (nearly all schoolmasters) listed in the AIGT's first annual report,[102] 52 were not members of the LMS but nevertheless were ready to join an organisation that addressed wider mathematical issues. The signal sent by the LMS in its response to Wilson's paper was that its interpretation of Rule 1, which stated that the objects of the Society were 'the promotion and extension of Mathematical knowledge', did not include debates about the teaching of mathematics. Rather, the Society wished to be a forum solely for the discussion of 'mathematical knowledge' itself, without considering how that knowledge was to be inculcated into the relatively unlearned. In this way, the Society

---

[n] Sylvester was in a slight muddle over Prospero's lines from *The Tempest*, V.1: 'I'll break my staff, / Bury it certain fathoms in the earth, / And deeper than did ever plummet sound / I'll drown my book.'

sought to rise above controversy by ring-fencing what it perceived as issues concerning 'mathematical knowledge' from peripheral issues with which it wanted nothing to do.

The desire to avoid controversy derived from a long-held belief that pure mathematical knowledge was so certain that no dissent or debate concerning its truth was possible. In 1815 Charles Hutton, whose career had been spent at the Royal Military Academy, expressed it thus:

> Pure mathematics has one peculiar advantage, that it occasions no contests among wrangling disputants, as is the case in other branches of knowledge: and the reason is, because the definitions of the terms are premised, and every person that reads a proposition has the same idea of every part of it. Hence it is easy to put an end to all mathematical controversies, by showing, either that our adversary is not constant with his definitions, or has not established the true premises, or that he has drawn false conclusions from true principles; and in case we are not able to do either of these, we must acknowledge the truth of what he has proved.[103]

In 1882 this view was updated, albeit with slightly less confidence, by Samuel Roberts, when – clearly disturbed by the advent of non-Euclidean geometry – he said in his presidential address that:

> On rare occasions, dissensions of a metaphysical order are historically prominent within the technical domain, but the citadels of pure science [pure mathematics] have usually been respected. ... Misgivings as to the permanence of our repose have been occasioned by modern speculations, which have culminated in the doctrines of hyperbolic, elliptic, and n-dimensional spaces.[104]

However, what he referred to elsewhere in his address as the 'instinctive prudence' of mathematicians governed not only the whole operation of the Society, but also the members' choice of subject matter. Although algebra became their preoccupation at the expense of geometry, it was usually of a particularly British kind that had been developed by Cayley and Sylvester as the theory of invariants, and its devotees (referred to as 'l'Eglise Invariantive', according to Sylvester)[105] pursued their researches along lines very different from those taken by Continental workers, for whom senior British pure mathematicians would not have made natural bedfellows when it came to publication. There was no tendency, as might have been expected if the Society was becoming more 'international', for it to attract a larger proportion of overseas contributions as the century progressed; in fact the converse is true, and in the 1890s the honorary foreign members contributed nothing at all; they were obliged only to receive a diploma with good grace, and clearly felt no particular urge to favour the *Proceedings* with their papers.[106] Thus the *Proceedings* was ineffective as a means of keeping British mathematicians abreast of developments on the Continent, and members who relied on it for this purpose would have had a very one-sided view of mathematical progress.

Although for most members the *Proceedings* formed their main link with the Society, for a few it was the monthly meetings that were the focal point. These were held in the Society's rooms at eight in the evening, following a Council meeting, until in 1900 they were moved to five in the afternoon.[107] Of course, only members resident in London or its suburbs could normally attend, and although for many years Cayley and Glaisher regularly travelled up from Cambridge, none of the other Cambridge members did: Glaisher commented that, in addition to himself, 'the regular habitués in the early days were Cayley, Smith, Clifford, Hirst, Roberts, Merrifield, and Walker'.[108] Henry Smith was based in Oxford, but the last five all lived in London. Thus in a Society of well over 100 members, the monthly meetings were regularly attended by only eight or so (along with an occasional guest), and the attendance of more than a dozen was rare.

By the mid 1870s the Society had acquired the form and characteristics that it was to retain for the rest of the century. Notwithstanding its name, it was clearly a 'national' society, possessed a straightforward set of rules, and was run in a gentlemanly, non-confrontational manner that avoided what it saw as peripheral issues. The Wilson affair had shown that the mood of the Society was to maintain the status quo, a sentiment expressed in 1882 by Samuel Roberts in his presidential address:

> Within the borders of our science, a neutral plateau is found, where minds of different philosophical temperament can harmoniously cooperate. This neutrality is precious, and should not be compromised by the intrusion of controversies, which

have ever been active at the outskirts, into interior regions hitherto uninvaded.[109]

This focussing of attention solely on the dissemination of 'mathematical knowledge' meant that the Society devoted none of its resources to education, or the support of students, or the encouragement of mathematical research; nor did it try to generate any public interest or discussion. Indeed, from about 1875 onwards the Society entered a period of stasis until the end of the century, with very little to disturb the even tenor of its way. Individual members, of course, had their achievements, their successes and failures, but the Society as such concentrated on holding its monthly meetings and publishing the *Proceedings*. There were, however, two duties that it carried out from time to time, and that had the potential to raise its international profile and encourage international cooperation. These duties were the election of honorary foreign members, and the triennial award of the De Morgan medal.

After Michel Chasles became the first honorary foreign member in 1867, no more were appointed until 1871, when the Council, hoping that the status of the *Proceedings* would be enhanced by contributions, resolved that 'Though it was not considered desirable that any paper should be demanded from these gentlemen, yet it was considered fitting to apprize them that the Society would be happy to receive such at their hands.'[110] However, as we have seen, the Society was to be disappointed, and very few papers resulted. The procedure for electing honorary foreign members was devised ad hoc on each occasion, and was often concerned with the nationality of the candidates, as much as with their mathematical attainments; thus in 1874 'it was thought that two German mathematicians should be elected'.[111] This desire to bring lustre to the Society by displaying broad international connections was shown most clearly in 1878, when it was decided to elect five new honorary foreign members, with one being elected in an 'open' category, and one from each of France, Germany, Italy and Scandinavia.[112] By the end of the century the Society had elected as honorary foreign members 22 of the most eminent foreign mathematicians, but the benefit generated was minimal.

Perhaps the most high-profile opportunity to promote the Society as 'international' came with the triennial awards of the De Morgan medal. A few years after accepting this responsibility the Society gave an indication of the importance that it attached to the award by losing the die, and 'After a long-running search for the missing

De M. medal, the estimate for making another one was 70 guineas',[113] although the eventual cost was only £10.[114] The Council's discussions of the award are characterised by a lack of enthusiasm that contrasts markedly with the hopes of Samuel Roberts; although he was concerned to avoid controversy, he also wanted to promote new work and widen the geographical reach of the Society, for in his presidential address he said:

> ... having regard to the cosmopolitan character of our science, and to the great value which must attach to our cordial co-operation with foreign men of science, the award should be made without regard to nationality. ... I confidently believe that this distinction, modest though it may be in material value, will be regarded by its recipients as a great encouragement.[115]

This statement, with its use of the words 'foreign' and 'encouragement,' carried the clear implication that the recipients should be up-and-coming mathematicians from overseas, a circumstance that would undoubtedly help to foster international relations. However, when in 1884 the Council first discussed the award, 'the only names suggested were those of Prof. Cayley & Prof. Sylvester – the feeling being in favour of the former, supposing no objection on the ground of the former being a member of the council'.[116] At the next meeting 'Prof. Cayley mentioned that he intended to propose Capt. Macmahon for the De M. medal. A card from Dr. Hirst proposed Herr Schubert.'[117] Captain (later Major) Percy MacMahon worked in the field of invariant theory, which was substantially the creation of Cayley and Sylvester, and at that time he was a 27-year-old instructor in mathematics at the Royal Military Academy, Woolwich. The nomination of Schubert, who was aged 35 and a schoolteacher in Hamburg, was the only one on the lines suggested by Roberts, which led Hirst – referring to the nominations of Cayley and Sylvester – to write in his journal: 'My contention was that the Medal would be more appropriately bestowed on a younger and a rising mathematician.'[118] Cayley and Schubert were finally put forward, but Hirst then withdrew Schubert's name, explaining: 'I did so after having written to Cayley who seconded my proposition on the 8th of May. I had hoped that Cayley would have declined the Medal; but he did not do so, and as I did not wish to divide the Council on the subject, I resolved to withdraw Schubert. It was awarded to Cayley today.'[119] It is hardly surprising that 'Mr. Glaisher proposed at December to suggest a change in the arrangements for the award of [the] De Morgan medal.'[120]

Whatever these new arrangements may have been, they had little effect. When in 1887 the Council next came to consider the medal, 'a desultory conversation ensued in the course of which no name besides that of Prof. Sylvester was brought forward'.[121] So that settled it: in June, as the only nominee, he was unanimously elected.[122] Indeed, the tendency to favour senior British members of the Society continued until the end of the century, because of the six recipients during that period, only one – the distinguished German mathematician Felix Klein, in 1893 – was from overseas; but he certainly did not need 'encouragement' from the LMS, the giving of which (Samuel Roberts had suggested) should have been one of the purposes of the medal. And as if to atone for this lapse from insularity, in 1896 the Society made its most inexplicable award, somewhat ironically to Samuel Roberts himself. Although second only to Cayley as a prolific contributor to the *Proceedings*, he was not considered at the time, nor subsequently, to be a mathematician of any particular distinction. Roberts seems to have been aware of this, for his valedictory address included what sounds suspiciously like an apologia for a society of second-rate mathematicians:

> The residue of permanent doctrine, of theorems and formulæ, which confer an immortality, very imperfectly represents what we owe to our predecessors. Their unremembered work was justified by suggestiveness and originality, which stimulated the minds of contemporaries, by aid rendered to new comers, destined perhaps to eclipse their masters, and by the ever-growing generalisation of ideas leading up to more comprehensive theories. In estimating our usefulness, gentlemen, we are entitled to take account of considerations of this kind.[123]

This may well have been true, but it is difficult to imagine Cayley or Sylvester presenting the membership in such a light, and suggests that Roberts believed that he himself would go down in history as the author of 'unremembered work'. Furthermore, for the De Morgan medal he was in competition with six other nominees, including much younger European luminaries such as Poincaré, Hilbert, Boltzmann and Mittag-Leffler,[124] the value of whose work was already clear, and whose names were later to become some of the most distinguished in their fields. The award seems to have generated no little heat in Council, because the original minutes of the discussion in May were cut out of the minute book and rewritten. However, in June Roberts was nevertheless elected the medallist.[125]

The tendency of the Society to avoid engagement in other activities, even when doing so would have helped to raise its profile overseas, is evident in its attitude towards two international congresses of mathematicians that were held before the turn of the century. The first, in Chicago in 1893, was associated with the World's Columbian Exposition, and was the subject of an attempt by the German government to establish – to an even greater extent than it already existed – the idea that American postgraduates seeking to continue their education by visiting Europe should naturally gravitate towards Germany. The Germans despatched to the congress Felix Klein, probably their most renowned mathematician, and he invited delegates 'to visit the German University Exhibit at the World's Columbian Exposition, and attend his exhibition and explanation of mathematical models° and apparatus.'[126] Klein was an unofficial guest of honour, and a vote of thanks was tendered to him 'for his very valuable contributions to the proceedings of the Congress and for his interesting expositions of the mathematical material in the German University Exhibit at the Exposition'.[127] Klein was building on foundations that had already been laid, for William Story, the president of the congress and himself a member of the LMS, eagerly acknowledged 'the indebtedness of American mathematics in general to the influence and inspiration of German mathematicians'.[128] The comments of Thomas Fiske in 1904 show the extent of Germany's success:

> The lectures of Professor Klein were in particular the Mecca sought by young Americans in search of mathematical knowledge. I think that it may be said safely that at present ten per cent of the members of the American Mathematical Society have received the doctorate from German universities, and that twenty per cent of its members have for some time at least pursued mathematical studies in Germany. It is not surprising that as a result a large portion of the American mathematical output shows evidence of German influence, if not of direct German inspiration.[129]

° Usually, the term 'mathematical model' should be taken as meaning the theoretical representation of natural or experimental processes, using mathematical concepts and systems of equations. However, sometimes (as in this case) the term can refer to physical models of geometrical concepts such as surfaces or three-dimensional figures.

For many Americans the attraction of Germany, apart from the charismatic Felix Klein, was that it offered research degrees, which Britain did not, and this would have been made very clear to those who attended the congress.[130] Of course, given the travelling involved, it was not expected that European mathematicians would attend the congress in droves, and only three did so other than Klein: another German, an Austrian and an Italian. However, when it came to submitting papers the Europeans spared no effort and contributed 26, twice as many as their American counterparts; and although 16 of the European papers came from Germany, 10 came from elsewhere. However, the number of papers from British mathematicians was zero.[131] For them, and for the LMS, the congress might just as well never have occurred – it was not even discussed in Council.

Similarly, at the congress held in Zurich in 1897 there were no papers from British mathematicians, although it was attended by three Britons: Ernest William Hobson and Joseph Larmor from Cambridge, and John Mackay from Edinburgh.[132] Hobson and Larmor were both members of the LMS, but there was no suggestion that a report on the congress in the *Proceedings* might have been of interest. The only comment was by the Society's president, E B Elliott, when he excused the lack of participation at Zurich by explaining that it clashed with a British Association meeting in Canada. He added: 'On another occasion, we trust that English mathematicians will be numerously, as well as strongly, represented.'[133]

The Society did lend two of its mathematical models to the Munich Exhibition of Mathematical Models, Apparatus and Instruments in 1893,[134] but this was largely due to the initiative of Olaus Henrici, who was born in Meldorf, which is now in Germany but was then Danish. His education had been at Karlsruhe, Heidelberg and Berlin, studying under some of the finest Continental mathematicians such as Hesse, Weierstrass and Kronecker. At the age of 25 he was unable to find suitable employment in Germany and so moved to Britain, which was to remain his home for the rest of his life. He was therefore unique in being a first-rank mathematician from the Continent who became absorbed into the British mathematical scene:[p] he held various professorships in London, was elected Fellow of the Royal

---

[p] Hirst had received his mathematical education on the Continent but was a Yorkshireman by birth.

Society in 1874 and was president of the LMS for two years from 1882. It follows that the loan to the Munich Exhibition, which was undertaken at his behest, is not evidence that the Society was becoming more international in outlook, for he was Continental European by birth and education, and so completely unrepresentative of the Society's members.

The foregoing discussion shows that the Society had no direct participation in international mathematical affairs up to 1900. Therefore a presentation of the Society as having become 'international' during that period cannot be sustained, because it could hardly have done less to promote and encourage the cause of internationalism. It mattered not whether that internationalism referred to Europe, which is where most new mathematics was arising, or the United States, which was just beginning to make its mark in mathematical research, or the Empire, which was the focus of interest in international affairs for most British people; the Society regarded them all with equal indifference.

On the home front, there were two attempts to acquire a royal charter, in 1871 and 1886, but on both occasions the Society failed to gain the approval of the Privy Council.[135] However, although it eventually became clear that the Society would not receive a royal charter in the foreseeable future – it had to wait until its centenary year of 1965 – some form of incorporation was certainly desirable, for the ambiguous legal status of unincorporated learned and scientific societies might have caused problems in any litigation.[136] Therefore in 1894 the LMS incorporated under the Companies Act 1862 as a company limited by guarantee, although it was allowed to omit the word 'Limited' from its name, so there was no apparent change in its status.[137] This form of incorporation was not remotely comparable to incorporation by royal charter because it said nothing about the standing of the LMS as a learned or scientific society, but simply enabled it to 'adopt the complete organisation of a corporate company'.[138] The Society's new legal status, which in an insolvency limited a member's liability to one guinea,[139] had no significant consequences, although honorary foreign members should thenceforward have been styled honorary associates because they were not legally members of the company.[140]

The change in the Society's legal status heralded other changes, slower and less well-defined, that began in the new century. Most of the early members, such as Augustus De Morgan, Cayley and Sylvester, had been educated in the Cambridge tradition of an overarching natural philosophy that held sway in the first half of the nineteenth century, but they were succeeded by a generation for whom the sciences had become increasingly specialised and fragmented. This was also the era of the rising professions, and consequently users of mathematics such as physicists, chemists, civil engineers, astronomers, actuaries, and those in many other occupations all benefitted from professional or learned societies that catered for their specific needs and interests, and that published their own journals. Only the pure mathematicians had nowhere to go other than the LMS, and there they kept alive a culture in which pure mathematics offered certain knowledge that was of a different order from the disputatious knowledge offered by the sciences. As a result, the Society became one in which a pure mathematician would feel particularly at home; so although in the *Proceedings* the number of papers in applied mathematics increased in line with the membership throughout the Victorian period, there was in the new century a slow but sure diminution in the overall attractiveness of the Society for applied mathematicians and mathematical physicists. In 1916 this was remarked upon by the president, Joseph Larmor, in his valedictory speech:

> Of recent years the question must have presented itself to not a few of our authors whether the *Proceedings*, developing in so abstract a direction, are now quite as suitable a place for the publication of mathematical physics as they were in the days when Maxwell and Kelvin, and Rayleigh and Routh, were frequent contributors.[141]

Although Larmor was engaged in wishful thinking when he said that Kelvin was a frequent contributor – for he contributed only one paper, on the theory of sound – the trend to which he referred was clearly evident, and would increase during the subsequent decade.

In the new century the LMS continued to grow, but at a slower rate than hitherto: by 1919 its ordinary membership had reached only 312, a rate of increase over the two decades that was half that of the 1890s. This slowdown was not an effect of the First World War, but is inevitable for any new society with a limited pool of potential members; yet that this happened with such a low membership emphasised the Society's specialised and elite nature. Furthermore, there was only a modest increase in the number of honorary associates, and as in the previous century they failed to support the *Proceedings*, for the only contributions during the 30 years from 1890 to 1919 came in a flurry of activity in the period 1900 to 1903, when five of them contributed a total of 67 pages, mostly from Poincaré and Hilbert.

As the Society matured it became more difficult to judge whether or not it was meeting the expectations of its members because about half of them had taken out life subscriptions, which offered good value for anyone in robust health who anticipated continued membership.[142] An annual member could resign or allow his membership to lapse if he felt that it no longer warranted a guinea a year – in 1891 Karl Pearson had been a high-profile example – but life members were always on the books; if Pearson had been one of them we would never have known that he had no further interest in membership.[143] Therefore it cannot be argued that a membership list bolstered by so many life members offered support for the increasingly pure direction in which the Society was heading.

This was a period of rapid growth in the provision of higher education in England and Wales, and during the second half of the nineteenth century many new colleges were founded at which students could study for University of London degrees; some of these colleges later became universities in their own right. So by 1910, in addition to the universities that had existed in 1890 (Oxford, Cambridge, Durham, London and Victoria), the Universities of Wales (1893), Birmingham (1900), Liverpool (1903), Leeds (1904), Manchester (1904), Sheffield (1905) and Bristol (1909) had received their royal charters in the years indicated. Liverpool, Leeds and Manchester had previously been the colleges of the federal Victoria University, which was effectively dissolved when Liverpool and Manchester sought autonomous status. However, these additional educational opportunities did not significantly increase the pool of potential members of the LMS, for the standard of entrants to the new colleges and universities was low, and many of the courses did not lead to a degree.[144] Such institutions were often very dependent on funding from local sponsors, and consequently had to gear their courses to the disciplines that those sponsors deemed to be of value.[145] This meant a strong emphasis on the teaching of practical subjects, with any research being heavily biased towards applications that would be relevant to commerce and industry; for example, at Bangor in Wales the mathematics department 'from 1910 focused much of its activity on aeronautics, not only studying the theoretical aspects of stability but actually building and flying a "flying machine" in 1911'.[146] Given the extent to which most of the new institutions had to direct much of their effort towards providing a more elementary level of instruction for the local population, either as an end in itself or as a preliminary to an undergraduate course, it is unsurprising that there was little enthusiasm for the LMS amongst staff, graduates or students: by 1919 the Society had just one member whose principal university degree was from Liverpool, two members from Manchester and one from Wales – a total of four from the seven new universities listed above. There was also one member with a Durham degree.

From this, it might be inferred that the influence of Cambridge continued unabated, and that inference is perfectly justified, for there was only a small reduction in the proportion of members who were Cambridge graduates. Their continuing dominance in the affairs of the Society is seen particularly in the composition of its Council, which took all decisions concerning policy, planning and administration, and was ultimately responsible for the content of the *Proceedings*. Throughout the period from 1900 to 1919 the Council consisted of the president, two or three vice-presidents, a treasurer, two secretaries, and nine or ten other members, the total always being 16. Thus over the 20-year period there were 320 slots available, of which Cambridge graduates occupied 244 (76%), London graduates 12, and Oxford graduates 42. The remaining 22 slots were occupied by Allan Cunningham and Percy MacMahon, two army officers who had not graduated from any university, although MacMahon was awarded an honorary DSc by Cambridge in 1904. Throughout the period both the secretaries held Cambridge degrees, and in 1919 J E Campbell became the first president for more than 20 years to have a degree from elsewhere: he graduated from Oxford. All the Council members with Cambridge degrees were wranglers, and they had an average position of 3.4 in the order of merit. With regard to occupations, 83 per cent of the Council slots were filled by academics. Finally, more than 80 per cent of the pages in the *Proceedings* were contributed by Cambridge-educated members.

There was a dearth of papers from the Empire, the contributions during the entire two decades being from Australia (24 pages), Canada (4 pages), and India (12 pages). This is not through a lack of ordinary members in these three countries, which contributed more than 7 per cent of the membership during the period. However, British graduates who went abroad normally did so in an official capacity, and so there would have been little opportunity for indulging in any mathematical interests. This is apparent from comments by S M Jacob, a Society member and old India hand, showing the disfavour with which a mathematical education was regarded. After noting that there were only about five positions in the Indian civil service that required significant mathematical ability, and that a trained mathematician occupied only one of these, he continued:

> All the other mathematicians were absorbed into routine administrative work and rarely had anything mathematically more complicated than working out a percentage: in fact, any attempt at mathematical reasoning of any complexity would have aroused patronising comment from their classical colleagues in the Indian Civil Service, who outnumbered the mathematical by 20 to 1. If, perchance, a mathematician had come from a 'good' (I had almost written 'approved') school, like Eton, Harrow, Winchester or Charterhouse, he might have reached to a Governorship; but I cannot remember such a happening in the Punjab in the first 25 years of the present [twentieth] century.[147]

An 'approved school' was formerly a government-run institution for juvenile delinquents. Jacob's message is clear and helps to explain why the Empire's contribution to the *Proceedings* was so low; consequently, the bulk of the papers was at all times contributed by authors resident in Britain.

In the first two decades of the twentieth century, the Society's central activity continued to be the publication of members' papers in the *Proceedings*. However, during that period only 116 members availed themselves of this facility, out of 443 who could have done so. Although some members may have published elsewhere, most were simply passive, notwithstanding that by 1919 almost three-quarters of them held academic positions; so we have to assume that the main benefit enjoyed by members lay in reading what others had published, rather than in publishing their own work. However, there could also have been another, less obvious, motive for membership, and that was to enjoy the illusion of communing with the great: a former wrangler, isolated perhaps in a country rectory, might believe it worth a guinea a year to be able to say to himself, even if to no-one else, that he was a member of the same society as were mathematicians such as Bertrand Russell, Hardy and Littlewood, or honorary associates such as Poincaré and Hilbert, and so was on terms of fellowship with them.[q] Membership of the Society was not just a matter of fact but also a matter of pride, an assertion borne out by the frequency with which such membership is mentioned in biographical entries, obituaries, and on the title pages of books. In 1881 W Marsham Adams had gone further, and described himself in his *Shilling Manual of Trigonometry* as 'formerly Member of the London Mathematical Society', presumably believing that the honour of having been thought worthy of membership outweighed the suggestion of penury; he had not been a member since 1876.[148] However, he later gave himself over to writing on the mysteries of Egyptology, so perhaps the Society was proving surplus to requirements.

The business of publishing papers was causing increasing problems for the Society, both financially and logistically. The *Proceedings* had no formal editor, and although the Council took final decisions as to what should be published, it rarely overrode the opinions of the referees. However, the production of the *Proceedings* was under the joint control of the secretaries,[149] and this became increasingly problematic as the complexity of papers increased and authors preferred to deal directly with the printer. Mathematical typesetting was far more time-consuming than standard typesetting, and this caused difficulties not only for the Society but also for any publisher of mathematical papers.[150] Although the Society might have been expected to lead the way in any move to standardise mathematical typography, as usual it was reactive rather than proactive, and it was left to the Royal Society, in 1906, to enquire whether the LMS would 'be willing to cooperate ... in an endeavour to secure greater uniformity in mathematical printing'.[151] The Council of the LMS appointed a committee to consider the matter, but it appears never to have met, and the Council merely advised the Royal Society that it 'would be prepared to cooperate if desired'.[152]

After this lukewarm response nothing transpired until 1909, when the Royal Society issued a circular, *Practical Suggestions on Mathematical Notation and Printing*, and the LMS agreed to distribute it with the *Proceedings* and provide new members with a copy.[153] This

---

[q] They had been elected honorary associates in the new century.

represented the first move by the Society to become something more than a free association of individuals, for it was here attempting to exert some sort of authority over its members in standardising an aspect of mathematical practice. However, the results cannot have been encouraging, for a few years later the Society enlisted the aid of its printer to produce *Suggestions for Notation and Printing*, which was issued with the *Proceedings* in 1914; it contained an instructive example of the complex work with leads, spaces and quadrats that was required to cope with the various sizes and alignments of mathematical symbols.[154]

The attempt by the Society in 1909 to influence mathematical practice was followed by another, for it was not merely typography that was causing problems. Authors, eager to secure priority, were sometimes rushing into print too soon, and this circumstance, along with the increasing length and complexity of mathematical papers, led to frequent corrections to the final manuscript after it had been submitted. These generated significant additional printing costs, for once the type had been set up any changes could require the resetting of all subsequent pages. It was Hardy who found the solution: in 1910 'three letters were read from Mr. Hardy raising the question of the desirability of publishing Abstracts of work in progress by which Abstracts authors might secure priority without being forced into premature publication of papers'.[155] A committee was set up to consider all means whereby the printing and publishing of papers might be expedited, and it made a number of recommendations, including the adoption of Hardy's proposal. The Society's view was contained in a draft circular to authors:

> It sometimes happens that, in a subject that is growing rapidly, an author is hurried into writing a paper in order to secure priority before he has thought out the matter fully, and afterwards has to write another to supersede some of his previous work. By the plan of publishing abstracts he can gain time for more mature consideration without losing priority. An abstract of a paper which is not intended to be published, but to mark a stage of progress, may need to be rather longer than an abstract of finished work. On the other hand, an abstract which is intended to announce a result without any indication of proof may be quite short without losing in value.[156]

Although the *Proceedings* thenceforward started to contain abstracts, the standard of authors' manuscripts continued to be a cause for concern, and in 1918 'Mr. Hardy proposed that a printed form of acceptance should be sent to authors' along with the circulars on printing, 'with the object of inducing authors to revise their papers carefully before sending the MSS. to press'.[157] In this way the Society was seeking to reinforce its earlier message that authors could benefit from the publication of interim results rather than completed papers, and formal expression was given to an extended role for the Society: hitherto its purpose had been to disseminate mathematical knowledge, but now there was an additional function, which was of value to the author rather than the reader, namely, to establish priority in mathematical research.

While the duty of publishing the *Proceedings* was inviolate, financial considerations continued to govern the Society's attitude towards other activities. As in the previous century, the cost of printing was the major outgoing, typically absorbing about 85 per cent of the Society's income; the rent of the rooms was the only other significant cost. However, the printer's bill fluctuated wildly from year to year, being largely determined by the number and length of the papers submitted. Consequently, budgeting was difficult and the Society lived in a constant state of financial uncertainty as to whether it would have sufficient resources to pay the printer. The situation was not helped by the Society's failure to account properly for accrued but unpaid expenses, which made it impossible to judge its true financial state until a move to an accruals basis of accounting in 1913 converted overnight an apparent accumulated surplus of £42 into an accumulated deficit of £38. After the war the increased cost of publication of the *Proceedings* imposed such a burden that in 1919 and 1920 the Society appealed to the Royal Society for a grant of £100 in each year.[158]

It is therefore understandable that the Council was always wary of any enterprise that might involve expenditure. For example, in 1900 A E H Love reported that the Society had been asked to assist in the production of the proposed International Catalogue of Scientific Periodicals, and that he had 'replied that on a previous occasion when similar enquiries had been made it had been pointed out that the difficulty of the expense might be serious as our Society is not a rich one but that I hoped this would be overcome & that our Society would be able to do what was asked of it'.[159] And when a similar enterprise was mooted in 1918, the Society gave a similar response.[160] However, this was a reflection of the Society's priorities: if it had been more discriminating in what it published in the *Proceedings*, then the reduction in printing costs would have made funds available for use elsewhere. As it was, the Society showed no inclination to be involved with other societies or institutions, and its concentration on publishing shows the extent to which it valued the nature of pure mathematics, which was now irrevocably isolated from the applied sciences.

It is apparent that from its earliest days the Society always interpreted rigidly its Rule 1, which set out as the object of the Society 'the promotion and extension of Mathematical knowledge', and thus avoided divisive pedagogical issues. Throughout the nineteenth century its outlook had remained that of a severely elite organisation, never making concessions in an attempt to popularise the subject or attract student members. This attitude was carried over into the new century: for example, in 1906 'a letter was read from the Teacher's Guild asking if the Society would take part in an Educational Congress. It was agreed that the proposed congress is outside the range of the Society's activity.'[161] Then in 1910 'letters were read from Mr. T. J. Garstang asking the Society to take some action to counteract a Circular of the Board of Education in regard to the teaching of Elementary Geometry. It was proposed ... that a Committee be appointed to report to the Council on the best method of teaching elementary geometry.'[162] Garstang was an Oxford graduate and schoolmaster who had joined the Society in 1908, and the object of his disapproval was Circular 711, 'Teaching of Geometry and Graphic Algebra in Secondary Schools', issued in March 1909, which charted a course between a slavish adherence to Euclid and an undue reliance on practical work. Given the background of the members of the LMS it was probably the tendency to diminish the role of Euclid that caused concern, fuelled by comments such as: 'Meanwhile, to continue to teach that all geometry rests on anything like Euclid's axioms is false as well as unnecessary.' It is clear that the opinion of the Society had not been sought on the question of how best to teach elementary geometry, and it appears that the proposed committee never met, let alone produced a report.

It was only in 1919 that increasing financial pressure forced the Society to canvass pedagogues, and even students, in an attempt to boost membership. Accordingly:

> a draft circular to Mathematical Departments in Universities and University Colleges, asking that the existence of the Society

should be brought to the notice of junior members of the staff and of promising graduates, was read. The circular was approved and it was agreed to send copies also to all members of the Society and to the editors of 'Nature' and of the 'Mathematical Gazette'.[163]

It will be recalled that the *Mathematical Gazette* was the journal of the Mathematical Association, which existed particularly for schoolteachers and originated with the Association for the Improvement of Geometrical Teaching. The choice of this journal, and of the popular general science publication *Nature*, suggests the extent to which financial considerations were forcing the Society to appeal to a new constituency in its quest for members. Furthermore, that the *existence* of the Society, by then more than 50 years old, needed to be conveyed to members of 'Mathematical Departments' is yet more evidence of its elitism, and the effect of its domination by Cambridge graduates.

On the positive side, this elitist and ascetic attitude also meant that the Society was not easily diverted by other issues, and early in its history Michel Chasles had been impressed by its concentration on mathematics as the sole criterion by which members were judged. Furthermore, women were admitted as soon as they could find sponsors, the first two being in 1881: Charlotte Angas Scott on 13 January and Christine Ladd Franklin on 12 May. More interesting, one might think, was the appointment of the first woman referee, which as far as can be determined occurred in 1899 when Frances Hardcastle (a member since 1896) was proposed for a paper by Dr Macaulay.[164] Women members could be regarded as potentially docile and submissive, but for a woman to referee a paper by a man implied a reversal of the usual Victorian gendering of authority and intellectual power.

As in the previous century, the Society did not participate in mathematical congresses, although it did agree 'to insert in the next issue of the Proceedings a leaflet calling attention to the Congress of Mathematicians to be held in Rome in 1908'.[165] More direct activity might have been expected for the congress held at Cambridge in 1912, of which Sir George Darwin (a member of the LMS) was president; but although the Society was 'anxious to do all in its power to help make the meeting in Cambridge a success',[166] it appears to have taken no further action. Doubtless many of its members were involved with the congress as individuals, so probably there was little need for additional input. Nevertheless, as with earlier congresses, the Society as such played no part.

It was during this period that Hardy and Littlewood established themselves on the world stage as being two of the foremost practitioners in mathematical analysis, which had superseded invariant theory as the principal interest of British mathematicians. Their work did not have the wide range and fundamental significance as that of the greatest Continental mathematicians such as Poincaré or Hilbert, and indeed mathematical analysis went rather out of vogue during the 1930s; but for a time it was in the forefront of pure mathematical activity, and Hardy, who surveyed his own achievements with unflinching objectivity, considered that at his peak he was the fifth best pure mathematician in the world, and that Littlewood was even better.[167] Hardy was also a keen, active member of the LMS, and as a tireless promoter of pure mathematics he was responsible for pushing the Society more firmly in that direction, something that no other pure mathematician of the time would have had the authority and charisma to do.

This period also saw Hardy and many other British mathematicians build up good relations with their German counterparts, relations that were to be adversely affected by the First World War. The Society was evidently slow to sever ties with Germany and its allies, for Alfred Basset complained that it was trading with the enemy; consequently 'the action of the Secretaries in stopping all Proceedings to Germany & Austria was reported and approved'.[168] By then Hardy was a leading light in the Society as well as in British mathematics, and after the war he did all in his power, both within and without the Society, to re-establish international scientific relations – but they had to be fully international. Thus, when in 1919 the Council of the LMS considered a proposal to form an International Union of Mathematical Societies, but excluding those of Germany and its allies, it was Hardy who moved that:

> the Council of the L.M.S. does not consider the organisation of an International Union of Mathematical Societies to be of urgent importance in the interests of mathematical research; and is of opinion that the formation of such a Union should be postponed until such time as it can be constituted on a fully international basis.[169]

Hardy's wording highlighted the extent to which British mathematicians in general, and he in particular, held their German counterparts in very high esteem; he wrote at the time that 'I have never valued any recognition more than what I have obtained in Germany; and I should regard the loss of my personal relationships with German mathematicians as an irretrievable calamity.'[170] The motion instigated considerable debate in Council, and was eventually passed with the omission of the words after 'postponed', a typical example of the way in which the Society would never commit itself to anything more than the minimum required to deal with the business in hand. In 1920 the International Research Council organised a committee to consider the whole question of the International Union of Mathematical Societies, and the LMS nominated three members, including Hardy; so despite the Society's history of non-involvement in international gatherings, it played its part in the slow process of bringing German mathematicians and scientists back into the fold.[171]

Hardy had come to dominate the Society: he took a deep interest in its affairs, and said in 1929 that since being appointed a secretary in 1917 he had neither missed a single word of a single paper, nor failed to attend a single Council meeting.[172] Largely through his own personality and determination, but also because he was rejected for war service on health grounds, he was able to extend and consolidate his influence at a time when most members were engaged in more pressing matters. However, he did not use that influence to widen the appeal of the Society or encourage it to be more receptive to applied mathematics or mathematical physics, notwithstanding the concerns that had been expressed by Joseph Larmor. In 1926 the *Proceedings* acquired a sibling, the *Journal of the London Mathematical Society*, which was destined to be almost exclusively devoted to pure mathematics, not least because other journals had become more attractive to applied mathematicians; on the probable continuance of this trend, Hardy commented: 'I will not pretend that the prospect causes me any particular distress.'[173] This was the period during which his internationally famous partnership with Littlewood was at its most productive, despite their being located in different universities due to Hardy's move to the Savilian chair at Oxford. He was Britain's best-known and most prolific pure mathematician, with a clear vision of the kind of Society that he wanted the LMS to be, a vision that did not embrace applied mathematics or mathematical physics.

Until the end of the First World War the Society continued in the same vein as in the previous century, holding its regular monthly meetings in no place other than London. One opportunity of making these meetings more social occasions, with the emphasis on the interchange of ideas, could have been taken in 1908 when the Cambridge mathematician Henry Frederick Baker arranged the provision of tea for members.[174] We saw earlier that at the end of the previous century the austere nature of the meetings, devoted as they were to the reading of long and complex papers, had caused concern to the president, Edwin Elliott. However, Baker's tea-based experiment in lightening the atmosphere does not seem to have been a success, for the next year the provision of tea was withdrawn.[175] Nevertheless this is not as trivial as it may at first appear, because Baker had great faith in the power of tea, and his probable intention was to effect a considerable change in the character of the meetings. This can be inferred from his success at Cambridge when, as the new Lowndean Professor in 1914, he instituted at his home Saturday-afternoon tea parties for his students and staff, where they could discuss their work and exchange ideas in a relatively relaxed setting. His earlier provision of tea at meetings of the LMS therefore suggests that he was hoping to encourage a similar atmosphere, and the failure of his efforts speaks of the Society's commitment to a rigid conception of the proper function of a learned society, and the strength of its resistance to change.

To conclude: in reviewing the Society's achievements during its first half century, it is helpful to consider the pointers that its first president, Augustus De Morgan, gave as to the direction that it should take. In his speech at the Society's meeting on 16 January 1865, he was clear that it should not become a problem-generating, problem-solving club, and in this respect his hopes were certainly realised. However, he also recommended that it should encourage a membership that was broadly based, both socially and intellectually, by setting its sights lower than a simple emulation of the Cambridge Philosophical Society, where everyone was (to put it bluntly) very clever; but as we have seen, this did not happen, and the membership – particularly the policy-making, paper-publishing membership – was increasingly dominated by wranglers, who had been educated and trained under a system that for many decades had included the passing on of mathematical methods and techniques like an heirloom. It is therefore unsurprising that the Society was adversely affected by the situation at Cambridge where, after a dalliance with abstract symbolism in the first half of the century, there was only a very slow absorption of ideas from Continental Europe.

That brings us to another part of De Morgan's speech, in which he emphasised the importance to the Society of studying the foundations of mathematics, and the connections between logic and mathematics. In the last quarter of the nineteenth century it was in these topics that some of the most interesting advances were being made, such as Cantor's famous and ingenious demonstration that there are 'more' real numbers than integers, and his consequent formulation of orders of infinity; Peano's definition of a completely counter-intuitive space-filling curve; and innovative work on mathematical logic undertaken by Dedekind, Frege and Peano. All this activity took place in Continental Europe, but as the Society had hardly any ordinary members there and attracted very few contributions from honorary associates, it was unable to present British mathematicians as participants; so it was left to Whitehead and Russell, early in the new century, to introduce to Britain ideas that were already revolutionising the foundations of mathematics. Thus British mathematicians were a quarter of a century behind the times when assimilating developments in those branches of mathematics that De Morgan had identified as being of particular significance and worthy of attention.

De Morgan's speech was not the only indicator of his hopes for the Society, for he also designed a badge that offered pure mathematics as transcending national, racial and religious boundaries. The Society failed to adopt the badge, quite understandably as it lacked any aesthetic appeal, but also showed no sign of recognising the message that it implicitly conveyed, which was to encourage harmony and cooperation between peoples and nations by appealing to the universality of mathematics; and it is difficult to see how this aim could have been realised without actively forging productive links with overseas mathematicians. However, the honorary associates brought little lustre – eminent mathematicians usually had several such honours – and the Society wasted the most high-profile opportunities to improve its international standing, because in the triennial awards of the De Morgan medal those who wished to honour younger, rising mathematicians from overseas were outvoted by those who wanted to reward senior members of the Society. This was symptomatic of a deep-seated insularity, whereas an outward-looking, contemporary approach could have monitored advances that were being made elsewhere, reinvigorated British mathematics, and generated public interest and discussion.

Furthermore, the Society's failure to recruit from new universities and colleges institutionalised the domination of British mathematics by Cambridge-educated mathematicians. This was not only a domination of the subject matter of mathematical research, for the highly competitive Mathematical Tripos examination was the most obvious manifestation of an ethos of personal struggle and achievement. This ethos was particularly required in invariant theory, as developed by Cayley and Sylvester, for a mathematician working in this branch of mathematics needed to undertake the algebraic manipulation of extremely long and complex expressions. The British had long prided themselves on producing mathematicians who were powerful calculators in algebra and arithmetic, as exemplified by the not-wholly-accurate but nevertheless record-breaking determination of π to 707 decimal places by William Shanks in 1873,[176] and Cayley and Sylvester had found the ideal branch of mathematics with which to indulge in this tradition. Thus there were two influences at work on British mathematicians – not just the intellectual isolation of being distanced from advances being made overseas, but also the sense that mathematics was an arena for the display of individual effort and prowess. The concept of the cooperative 'seminar' for mathematicians was late to arrive in Britain, and that this German word appeared in British English as an import from America, rather than direct from Germany, suggests the extent of British resistance to Continental influences.

So the verdict on whether the LMS was a force for good in presenting pure mathematicians as having some worthwhile role in society is mixed, at best. For them it certainly had value: the *Proceedings* kept them abreast of work that had been undertaken by their British confreres, and gave them publishing opportunities that they would probably not otherwise have enjoyed. Furthermore, the Society served as a model for several new mathematical societies overseas. However, it did little to engage with the non-mathematical universe, but rather immured itself in splendid isolation, keeping the world at bay as it preserved the engrained traditions of British mathematics. While it is usually unwise to speculate on what someone might have thought, it is difficult to avoid the conclusion that De Morgan would have been disappointed that such narrowness of view should have prevailed in the Society. Whether the Society would have been more valuable to its members, or more successful in public engagement, had it followed his prescriptions more closely, is another matter.

In the next chapter we will see how the circumstances that we have been considering affected the life and work of some prominent pure mathematicians of the time.

Notes and References

Many of the unreferenced claims and statements made in this chapter are derived from the author's statistical analysis of the London Mathematical Society, and of its *Proceedings*. Invaluable for the nineteenth century have been two publications issued by the Society in 1900, to celebrate its thirty-fifth anniversary. The first is a *Complete Index of all the Papers printed in the Proceedings of the London Mathematical Society (22 Albemarle Street, W.): Vols. I—XXX*, which indexed and classified the papers using a system that was based on that of the *Jahrbuch über die Fortschritte der Mathematik*. The second is an aggregated and corrected membership list, which has been taken as definitive concerning the membership during this period. For the twentieth century the author has used, amongst other sources, the *Jahrbuch*, Council Minutes, and the record of members' admissions published in the Society's *Proceedings*. To avoid repetition, we give now the archival references for the Minute Books in the archives of the London Mathematical Society. Council Minute Books: 1866–1872, LMSA 1/1; 1872–1885, LMSA 1/2; 1885–1892, LMSA 1/3; 1892–1894, LMSA 1/4; 1894–1916, LMSA 1/5. Council and Meetings Minute Books: 1916–1920, LMSA 1/6; 1921–1924, LMSA 1/7; 1924–1930, LMSA 1/8. Meetings Minute Books: 1865–1875, LMSA 2/1; 1875–1884, LMSA 2/2; 1884–1892, LMSA 2/3.

The original manuscript of the journal of Thomas Archer Hirst is now either lost or destroyed. However, a complete typescript copy is in the archives of the Royal Institution in London (RI MS JT/2/32) and the references below are to this.

1. Leone Levi, 'On the Progress of Learned Societies, illustrative of the Advancement of Science in the United Kingdom during the last Thirty Years', *Report of the Thirty-eighth Meeting of the British Association for the Advancement of Science* [1868] (London: John Murray, 1869), pp. 169–173, 197.
2. For a survey of the role of puzzles and problems in eighteenth- and nineteenth-century mathematical journals, see Sloan Evans Despeaux, 'Mathematical Questions: A Convergence of Mathematical Practices in British Journals of the Eighteenth and Nineteenth Centuries', *Revue d'Histoire des Mathématiques*, 20 (2014), 5–71.
3. Augustus De Morgan, *A Budget of Paradoxes* (London: Longmans, Green, 1872), p. 231.
4. De Morgan, *Budget*, p. 232.
5. G H Hardy, 'Prolegomena to a Chapter on Inequalities', *Journal of the London Mathematical Society*, 4, 1 (1929), 61–78, at 62.
6. Andrew Warwick, *Masters of Theory: Cambridge and the Rise of Mathematical Physics* (Chicago: University of Chicago Press, 2003), p. 254.
7. J W L Glaisher, 'The Mathematical Tripos', *Proceedings of the London Mathematical Society*, XVIII (1886), 4–38, at 7.
8. June Barrow-Green '"The advantage of proceeding from an author of some scientific reputation": Isaac Todhunter and his Mathematics Textbooks', in Jonathan Smith and Christopher Stray (eds.), *Teaching and Learning in Nineteenth-Century Cambridge* (Woodbridge: Boydell Press, 2001), pp. 177–203.

9. Letter dated 13 September 1935, Macmillan Archives (British Library), Add. 55197, fo. 127.
10. For a useful tabulated summary of European (including local British) societies, see Danny J Beckers, '"Untiring Labor Overcomes All!": The History of the Dutch Mathematical Society in Comparison to its Various Counterparts in Europe', *Historia Mathematica*, 28, 1 (2001), 31–47.
11. Charles Coulston Gillispie (ed.), *Dictionary of Scientific Biography*, V (New York: Scribner's, 1970), p. 414.
12. Hélène Gispert and Renate Tobies, 'A Comparative Study of the French and German Mathematical Societies before 1914', in Catherine Goldstein, Jeremy Gray and Jim Ritter (eds.), *Mathematical Europe: History, Myth, Identity* (Paris: Éditions de la Maison des Sciences de l'homme, 1996), pp. 407–430; and see this chapter generally for a comparative study of the founding and early years of the French and German societies.
13. Warwick, *Masters of Theory*, pp. 252–253.
14. J A Venn (comp.), *Alumni Cantabrigienses: A Biographical List of all known Students, Graduates, and Holders of Office at the University of Cambridge from the Earliest Times to 1900*, II, V (Cambridge: Cambridge University Press, 1940–1954), p. 247.
15. Sophia De Morgan, *Memoir of Augustus De Morgan* (London: Longmans, Green, 1882), p. 281.
16. De Morgan, *Memoir*, p. 281.
17. De Morgan, *Memoir*, p. 281.
18. Hirst's invitation is in the Archives of University College, London (MS Add. 69); George De Morgan's draft is given in De Morgan, *Memoir*, p. 282.
19. *List of Members* (London: London Mathematical Society, 1900), p. 1.
20. Thomas Archer Hirst, *Journal*, p. 1706.
21. Letter from George De Morgan to Ranyard dated 6 September 1864, quoted in *Proceedings of the London Mathematical Society*, XXVI (1894), 555.
22. Letter from Philip Magnus, quoted in *Proceedings of the London Mathematical Society*, XXVI (1894), 555.
23. *Proceedings of the London Mathematical Society*, XXVI (1894), 555.
24. *Proceedings of the London Mathematical Society*, III (1869), 233.
25. *Proceedings of the London Mathematical Society*, XXVI (1894), 555.
26. E F Collingwood, 'A Century of the London Mathematical Society', *Journal of the London Mathematical Society*, 41, 4 (1966), 577–594, at 577.
27. *Proceedings of the London Mathematical Society*, XXVI (1894), 555.
28. *The Spectator*, 13 May 1871, 3.
29. LMS Council Minutes, 10 June 1880.
30. LMS Council Minutes, 11 November 1880.
31. *Proceedings of the London Mathematical Society*, XXVI (1894), 555–556. See also XXVII (1895), 1, where Ranyard is referred to as 'joint founder of the old Society'.
32. LMS Council Minutes, 9 November 1871.
33. University College, London, *Calendar*, Sessions MDCCCLXV–LXVI, p. 153; MDCCCLXVI–LXVII, p. 177.
34. *The Times*, 18 February 1892, p. 5, col. E.
35. Venn, *Alumni Cantabrigienses*, II, II, pp. 275–276.
36. Letter dated 12 June [no year], London Mathematical Society Archive, University College London Library Services, Special Collections.
37. It is in the Archives of University College, London (MS Add. 69).
38. De Morgan, *Memoir*, p. 368.

39. The names provide the opening entry in the first General Minute Book, and are conveniently listed in Adrian C Rice, Robin J Wilson and J Helen Gardner, 'From Student Club to National Society: The Founding of the London Mathematical Society in 1865', *Historia Mathematica*, 22 (1995), 402–421, at 408.
40. All the quotations from De Morgan's address are from *Proceedings of the London Mathematical Society*, I, 1 (1865), 1–9.
41. Despeaux, 'Mathematical Questions', 58.
42. LMS General Minutes, 15 January 1866.
43. J W L Glaisher, 'Notes on the Early History of the Society', *Journal of The London Mathematical Society*, 1, 1 (1926), 51–64, at 61; and see the minutes of the Special General Meeting, 12 November 1868 in the General Minute Book. Glaisher also says that the date was changed to the second Thursday in each month, which is when the meetings were actually held, but the resolution itself gave the Council complete flexibility.
44. *Proceedings of the London Mathematical Society*, I, 2 (1865), 1–16.
45. *Proceedings of the London Mathematical Society*, II (1866), 2.
46. LMS General Minutes, 15 October 1866.
47. *Proceedings of the London Mathematical Society*, II (1866), 8, records the vote of thanks to him passed at the meeting on 24 January 1867.
48. LMS Council Minutes, 8 December 1870.
49. Letter to Robert Tucker dated 15 January 1891, loose in the LMS Council Minute Book at that date.
50. LMS Council Minutes, 11 October 1894.
51. LMS Council Minutes, 19 March 1866.
52. Sloan Evans Despeaux, 'Fit to print? Referee Reports on Mathematics for the Nineteenth-Century Journals of the Royal Society of London', *Notes and Records of the Royal Society*, 65 (2011), 233–252, at 246.
53. Rule 45 and Glaisher, 'Notes', 58.
54. LMS Council Minutes, 14 November 1867.
55. James Smith, *On the Quadrature of the Circle: Letter addressed to the Chairman and Members of the Committee of the Mathematical and Physical Section of the British Association for the Advancement of Science, at its 31st Annual Meeting, held in Manchester, 4th September, 1861* (Liverpool: Edward Howell, 1861), p. 15.
56. LMS Council Minutes, 9 November 1871.
57. LMS Council Minutes, 12 October 1893.
58. LMS Council Minutes, 16 April 1866.
59. LMS Council Minutes, 28 March 1867.
60. De Morgan, *Memoir*, p. 363.
61. Michel Chasles, *Rapport sur les Progrès de la Géométrie* (Paris: 'A l'Imprimerie Nationale', 1870), pp. 378–379.
62. 'Note Concerning the Mathematical Society of Paris', *Nouvelles Annales de Mathématiques*, second series, 11 (1872), 381–384.
63. 'Minutes of the Special Meeting on 15 Oct. 1866', *Proceedings of the London Mathematical Society*, I, 8 (1865), 14.
64. LMS Council Minutes, 8 January 1874.
65. LMS General Minutes, 10 December 1874; *Proceedings of the London Mathematical Society*, VI (1874), 20.
66. Adrian C Rice and Robin J Wilson, 'From National to International Society: The London Mathematical Society, 1867–1900', *Historia Mathematica*, 25 (1998), 185–217, at 196.

67. Glaisher, 'Notes', 52.
68. Glaisher, 'Notes', 57.
69. Colin Russell, *Science and Social Change, 1700–1900* (London: Macmillan, 1983), p. 224.
70. R V Wallis and P J Wallis, *Index of British Mathematicians, Part III: 1701–1800* (Newcastle upon Tyne: Project for Historical Biobibliography, 1993).
71. Glaisher, 'Notes', 61.
72. *Proceedings of the London Mathematical Society*, XVII (1885), 238.
73. E B Elliott, 'Some Secondary Needs and Opportunities of English Mathematicians', *Proceedings of the London Mathematical Society*, XXX (1898), 6–7, 10–11.
74. *Proceedings of the London Mathematical Society*, XXV (1893), p. 195.
75. LMS Council Minutes, 13 January 1876 and 11 May 1876; *Proceedings of the London Mathematical Society*, VII (1875), 241. See also Arthur Cayley, 'On Plücker's Models of certain Quartic Surfaces', *Proceedings of the London Mathematical Society*, III (1869), 281–285.
76. Glaisher, 'Notes', 62.
77. Jesper Lützen, 'International Participation in Liouville's *Journal de mathématiques pures et appliquées*', in Karen Hunger Parshall and Adrian C Rice (eds.), *Mathematics Unbound: The Evolution of an International Mathematical Research Community, 1800–1945* (Providence, RI: American Mathematical Society, 2002), pp. 91–92.
78. Warwick, *Masters of Theory*, p. 154.
79. Glaisher, 'Notes', 56.
80. LMS Council Minutes, 12 December 1872.
81. *Proceedings of the London Mathematical Society*, XXV (1893), 6–7.
82. Rice and Wilson, 'From National to International Society', 195.
83. *Proceedings of the London Mathematical Society*, XXVI (1894), 556.
84. University College, London, *Calendar*, Sessions MDCCCLXV–LXVI, p. 153.
85. LMS General Minutes, 15 January 1866.
86. *Proceedings of the London Mathematical Society*, I, 5 (1865), 10.
87. LMS General Minutes, 8 November 1866.
88. LMS General Minutes, 14 November 1867.
89. *Proceedings of the London Mathematical Society*, II (1866), 45.
90. LMS Council Minutes: 13 May 1869, 14 October 1869 and 13 October 1870.
91. LMS Council Minutes, 11 December 1873 and 8 January 1874.
92. LMS Council Minutes, 11 June 1874.
93. LMS Council Minutes, 8 May 1890.
94. Warwick, *Masters of Theory*, p. 188.
95. James M Wilson, 'The Early History of the Association: Or, the Passing of Euclid from our Schools and Universities, and how it came about. A Story of Fifty Years ago', *The Mathematical Gazette*, X (1921), 239–244, at 241.
96. *Proceedings of the London Mathematical Society*, II (1866), 69.
97. Wilson, 'Early History', 240.
98. A A Bourne, 'Further Reminiscences', *The Mathematical Gazette*, X (1920), 244–246, at 245. For a discussion of the Italians' experience, see Gert Schubring, 'Changing Cultural and Epistemological Views on Mathematics and different Institutional Contexts in Nineteenth-Century Europe', in Catherine Goldstein et al., *Mathematical Europe*, pp. 379–383.
99. *Proceedings of the London Mathematical Society*, II (1866), 69.
100. *First Annual Report of the Association for the Improvement of Geometrical Teaching*, January 1871.

101. James Joseph Sylvester, 'Presidential Address to Section "A" of the British Association', in *Report of the Thirty-eighth Meeting of the British Association for the Advancement of Science* [1868] (London: John Murray, 1869), pp. 1–9, at p. 6.
102. *First Annual Report of the Association for the Improvement of Geometrical Teaching*, January 1871.
103. Charles Hutton, *A Philosophical and Mathematical Dictionary: containing an Explanation of the Terms, and an Account of the several Subjects, comprised under the heads Mathematics, Astronomy, and Philosophy both Natural and Experimental; with an Historical Account of the Rise, Progress and Present State of these Sciences; also Memoirs of the Lives and Writings of the most eminent Authors, both ancient and modern, who by their Discoveries or Improvements have contributed to the Advance of them*, II (London: The Author et al., 1815), p. 23, Art. 'Mathematics'.
104. Samuel Roberts, 'Remarks on Mathematical Terminology, and the Philosophic Bearing of recent Mathematical Speculations concerning the Reality of Space', *Proceedings of the London Mathematical Society*, XIV (1882), 5–15, at 9.
105. James Joseph Sylvester, 'A Plea for the Mathematician, II', *Nature*, 1, 10 (6 January 1870), 261–263, at 262.
106. Rice and Wilson, 'From National to International Society', 207, say that in the 1890s 'Papers by foreign members such as Klein, Mannheim and Cremona are featured ... ', but this is wrong; the papers in fact appeared in the 1880s, which argues against the case for increasing internationalisation that they are trying to make.
107. Glaisher, 'Notes', 57.
108. Glaisher, 'Notes', 62.
109. Roberts, 'Remarks on Mathematical Terminology', 9.
110. LMS Council Minutes, 9 November 1871.
111. LMS Council Minutes, 10 December 1874.
112. LMS Council Minutes, 11 April 1878.
113. LMS Council Minutes, 14 May 1885.
114. LMS Council Minutes, 10 December 1885.
115. Roberts, 'Remarks on Mathematical Terminology', 6.
116. LMS Council Minutes, 13 March 1884.
117. LMS Council Minutes, 3 April 1884.
118. Hirst, *Journal*, 8 May 1884, fo. 2145.
119. Hirst, *Journal*, 12 June 1884, fo. 2148.
120. LMS Council Minutes, 12 June 1884.
121. LMS Council Minutes, 10 March 1887.
122. LMS Council Minutes, 9 June 1887.
123. Roberts, 'Remarks on Mathematical Terminology', 5.
124. LMS Council Minutes, 23 April 1896.
125. LMS Council Minutes, 11 June 1896, which is where the rewriting of the Minutes for May was authorised. See also the Minutes for 8 October 1896.
126. *Mathematical Papers read at the International Mathematical Congress held in connection with the World's Columbian Exposition, Chicago, 1893*, edited by the Committee of the Congress (New York: Macmillan, 1896), p. ix.
127. *International Mathematical Congress*, p. xii.
128. *International Mathematical Congress*, p. xii.
129. Thomas S Fiske, 'Mathematical Progress in America: Presidential Address delivered before the American Mathematical Society at its Eleventh Annual Meeting

December 29, 1904', *Bulletin of the American Mathematical Society*, XI, 5 (1905), 238–246, at 239–240.
130. The subject of research degrees is comprehensively treated in Renate Simpson, *How the PhD came to Britain: A Century of Struggle for Postgraduate Education* (Guildford: The Society for Research into Higher Education, 1983). The rise of American mathematical research is chronicled in Karen Hunger Parshall and David E Rowe, *The Emergence of the American Mathematical Research Community, 1876–1900: J J Sylvester, Felix Klein, and E H Moore* (Providence, RI: American Mathematical Society, 1994), where they also give a complete account of the Exposition.
131. *International Mathematical Congress*.
132. Ferdinand Rudio (ed.), *Verhandlungen Des Ersten Internationalen Mathematiker-Kongresses in Zürich vom 9. bis 11. August 1897* (Leipzig: Teubner, 1898), Verzeichnis der Teilnehmer, pp. 65–78.
133. Elliott, 'Some Secondary Needs and Opportunities', 8.
134. Walther von Dyck, *Katalog mathematischer und mathematisch-physikalischer Modelle, Apparate und Instrumente* (Munich: C Wolf, 1892), pp. 267 (No. 170), 283 (No. 199).
135. *Proceedings of the London Mathematical Society*, IV (1871), 146; XIX (1887), 2.
136. Henry Thring, *A Supplement to the Second Edition of the Law and Practice of Joint-Stock and other Public Companies; containing the Companies Act, 1867, with an Explanation and Notes, etc* (London: Stevens, 1889), p. 354.
137. The Companies Act, 1867, Section 23.
138. Thring, *Supplement*, p. 354.
139. Memorandum of Association, para. 7; the Memorandum and Articles of Association are reproduced in *Proceedings of the London Mathematical Society*, XXVI (1894), i–x.
140. LMS Council Minutes, 8 March 1894.
141. Joseph Larmor, 'Address by the Retiring President', *Proceedings of the London Mathematical Society*, 16, 1 (1917), 1–7, at 5.
142. *Proceedings of the London Mathematical Society*, XVII (1895), 1; XVIII (1896), 1.
143. LMS Council Minutes, 11 November 1891. A letter from the LMS asking him not to resign, and a further letter in 1897 asking him to consider renewing his membership, are in UCL Archives, Special Collections, ref. Pearson/11/1/19/60. He did not renew.
144. Gordon Roderick and Michael Stephens, 'Scientific Studies and Scientific Manpower in the English Civic Universities 1870–1914', *Science Studies*, 4, 1 (1974), 54.
145. Michael Sanderson, *The Universities and British Industry, 1850–1970* (London: Routledge, 1970), p. 29.
146. Sanderson, *The Universities and British Industry*, pp. 133–134.
147. S M Jacob, *Favour for Fools in a Decadent Empire: A Skeletal Autobiography* (H C Dunckley, 1972), p. 93.
148. W Marsham Adams, *The Shilling Manual of Trigonometry, illustrated by numerous Examples and Diagrams* (London: Burns and Oates, 1881).
149. LMS Council Minutes, 10 October 1901.
150. *The London Mathematical Society: Suggestions for Notation and Printing* (n.d., but received by the British Museum 27 August 1915).
151. LMS Council Minutes, 26 April 1906.
152. LMS Council Minutes, 14 June 1906.
153. LMS Council Minutes, 11 March 1909.

154. *Proceedings of the London Mathematical Society*, 14 (1914–1915), 1–4.
155. LMS Council Minutes, 8 December 1910.
156. LMS Council Minutes, 11 June 1911.
157. LMS Council Minutes, 14 March 1918.
158. LMS Council Minutes, 13 June 1918, 13 February 1919, 24 April 1919 and 23 April 1920.
159. LMS Council Minutes, 14 June 1900.
160. LMS Council Minutes, 14 March 1918.
161. LMS Council Minutes, 8 November 1906.
162. LMS Council Minutes, 28 April 1910.
163. LMS Council Minutes, 24 April 1919 and 15 May 1919.
164. LMS Council Minutes, 13 April 1899.
165. LMS Council Minutes, 14 March 1907.
166. LMS Council Minutes, 13 February 1908.
167. C P Snow, 'Foreword' to G H Hardy, *A Mathematician's Apology* (Cambridge: Cambridge University Press, 1967), p. 12; originally published (without the Foreword) 1940.
168. LMS Council Minutes, 12 November 1914.
169. LMS Council Minutes, 11 December 1919.
170. Letter to Gustav Mittag-Leffler dated 7 January 1919, quoted in Joseph W Dauben, 'Mathematicians and World War I: The International Diplomacy of G. H. Hardy and Gösta Mittag-Leffler as reflected in their Personal Correspondence', *Historia Mathematica*, 7, 3 (1980), 261–288, at 263.
171. LMS Council Minutes, 11 March 1920.
172. Hardy, 'Prolegomena', 61.
173. Hardy, 'Prolegomena', 62.
174. LMS Council Minutes, 30 April 1908 and 11 June 1908.
175. LMS Council Minutes, 13 May 1909.
176. William Shanks, 'On the Extension of the Numerical Value of π' and 'On Certain Discrepancies in the Published Numerical Value of π', *Proceedings of the Royal Society of London*, 21 (1872), 318–319, and 22 (1873), 45–46, respectively. In the latter paper he reported some minor transcription errors; the major errors were discovered by D F Ferguson: 'Evaluation of π. Are Shanks' Figures Correct?', *The Mathematical Gazette*, 30, 289 (May 1946), 89–90.

4 THE PURE MATHEMATICIAN AS HERO

The issues raised in the previous chapters concerning the relevance and status of pure mathematicians could blight the career and commitment to mathematics of even the most dedicated. Perhaps no one illustrates this better than James Whitbread Lee Glaisher, who was born in 1848 in Lewisham, south London, and spent the whole of his adult life living in rooms at Trinity College, Cambridge, where he had been admitted as an undergraduate in 1867; he died there after more than 60 years of residence in 1928. He had a particular interest in elliptic functions, a topic of great contemporary interest, but worked within a Cambridge tradition of real rather than complex analysis that soon rendered much of his work obsolete, and his completed treatises were never published.[a] Although until the end of his life he spent some time every day working at mathematics, from the turn of the century his particular interest was pottery.[1] He became a leading authority on the subject and built up a collection of great value which, together with £10,000 for maintenance and display, he bequeathed to the Fitzwilliam Museum, Cambridge, where it can still be seen.[2] His disillusion with his mathematical career resulted from the tension that existed between the demands of pure mathematics, as it was increasingly understood to be, and the outlook of a practitioner who fully supported the autonomy of

---

[a] Elliptic functions comprise a branch of analysis that has its origins in formulae associated with an ellipse. Real analysis uses the numbers ('real numbers') of everyday arithmetic; complex analysis supplements the real numbers by introducing a quantity represented by i, of which it is postulated that $i^2 = -1$. The numbers used in complex analysis are referred to as 'complex numbers'.

pure mathematics but was nevertheless unable to comprehend the extent and abstract nature of the most modern developments.

The problem was particularly acute for Glaisher because his father, also called James, was a traditional natural philosopher who, although without a degree, was a noted meteorologist, astronomer, balloonist, calculator of tables and stalwart of the British Association for the Advancement of Science.[3] There is much scope for confusion because the two of them were at the same time members of the London Mathematical Society, Fellows of the Royal Society and members of the British Association's Committee on Mathematical Tables. The father also enjoyed a particularly long and active life, having been born in 1809 and continuing with scientific work almost up to his death in 1903. Thus for 30 years their two careers ran in parallel, which is perhaps one reason why the younger James failed to embrace the emerging ideals of non-utilitarian pure mathematics with the same public enthusiasm as did more successful mathematicians such as Cayley and Sylvester.

As a boy, James junior showed signs of being set for a brilliant career. After St Paul's School he went up to Trinity to study mathematics, and when he was still an undergraduate his first paper (with which his father assisted) was communicated to the Royal Society by Cayley.[4] In 1871 he emerged from the Mathematical Tripos as second wrangler, having been coached by the very successful wrangler-maker Edward Routh,[5] and in the same year he obtained a fellowship of Trinity, a post as assistant tutor, and a joint-editorship of the *Messenger of Mathematics*. He also began to indulge his lifelong love of computation as a member of the British Association's Committee on Mathematical Tables: his colleague Andrew Russell Forsyth said of him that 'his devotion to numerical calculations for their own sake gave him inexpressible pleasure'.[6] In February 1872 he was elected a member of the LMS, in November of that year he joined its Council, and by the end of 1873 he had published more than 60 papers – only eight of them in the Society's *Proceedings*, but nearly half in the *Messenger*. Two years later, at the age of 26, he was elected Fellow of the Royal Society. Thus in the five years after graduation there were manifest the four parallel strands of his mathematical career – those of researcher, teacher, editor and committee member.

Glaisher was one of the first mathematicians at Cambridge to expound the value of pure-mathematical research at a time when it was hardly recognised in Britain. It was this concentration on obtaining new results, rather than in expounding things already known and understood, that accounted for his remarkable output as an author. However, Forsyth believed that Glaisher's work as a researcher had been blighted by the failure of mathematicians at Cambridge to follow many of the advances that had been made on the Continent, with the result that his talent was expended in:

> pursuing mainly well-established subjects by methods that were uninfluenced by the current developments of analysis then effected in France and Germany. ... Not a few of his early papers dealt with definite integrals. His results had a gravely limited significance, partly owing to the use of real variables only which belonged to the Cambridge habit, partly owing to the very narrow restricted methods, without any consideration of their functional bearing; he was satisfied with the methods that had proved fruitful under the genius of Euler and Legendre. Though not blind to the pitfalls inevitable in such a treatment of definite integrals, he made no excursions into the vast modern theory of integration which has revolutionized the subject out of all ancient recognition.[7]

Euler died in 1783, and Legendre in 1833. It is ironic, and illustrates how deeply embedded Cambridge traditions were, that Forsyth's own book *Theory of Functions of a Complex Variable*, published in 1893, has been criticised in a similar fashion by Leonard Roth, his last pupil, who wrote that Forsyth 'never comes within reach of comprehending what modern analysis is really about; indeed whole tracts of the book read as though they had been written by Euler'.[8]

Thus, although in Hardy's rather condescending estimation the best of Glaisher's work was 'really good',[9] his huge output, which in terms of rate of production at times rivalled Cayley's, had much less significance than its bulk would suggest. Cayley, who was held in considerable esteem by Continental mathematicians, had left Cambridge a few years after taking the Tripos in 1841 (when he was senior wrangler and first Smith's prizeman) in order to pursue a career as a conveyancing barrister, and did not return to academic life until he was recalled nearly 20 years later to take the new Sadleirian chair of pure mathematics. The researches that he undertook during his legal career, and for which he was widely honoured – in particular, the theory of invariants and the theory of matrices – and also the 300-odd papers that he published, arose from his leisure moments, of which he admittedly seems to have had plenty. Both Cayley and Glaisher had been through the Tripos mill, but Cayley's career choice had allowed him to develop his mathematics with an intellectual freedom that was much more difficult for Glaisher to obtain within the confines of Trinity.

Nevertheless, out of date though Glaisher's research methods may have been, as a teacher he was a resounding success, with elliptic functions being the subject that particularly fascinated him. Forsyth commented that 'wandered he never so far afield, he found himself returning time and time again to his beloved elliptic functions',[10] and Glaisher himself said that 'for fourteen years I have lectured regularly, each year, upon [elliptic functions], and no lectures of mine have been of so much interest to me'.[11] This enthusiasm infused his lectures at a time when elliptic functions were almost unknown at Cambridge, and Forsyth, whose praise of Glaisher the researcher seems more conventional than heartfelt, is unstinting in his praise of Glaisher the teacher:

> He was a member of the mathematical staff of Trinity College: as such, his lectures in courses on differential equations, or combination of observations, or elliptic functions, were a revelation to students. In those domains there were no recognised text-books in his earlier days: the examination schedule was not pedantically precise in its indicated range: and these subjects did not fall within the field of applied mathematics as favoured at Cambridge. So he was free to expound the subjects unfettered: this was done, systematically and continuously. The Tripos was not mentioned. His students were encouraged (then a rare practice) to consult original authorities; and, under his stimulus, they acquired a working knowledge of a subject without regard to the needs or the subtleties of examinations. The result was that successive bodies of Trinity students came under Glaisher's inspiring influence. In the course of the years, Cambridge pure mathematics went, in many directions, beyond the sphere of his own activities: such progress was due to men who, frequently, had been his students and had been encouraged by him in their research.[12]

Joseph John Thomson, who won the Nobel prize for physics in 1906 and was Master of Trinity when Glaisher died, had studied under him, and in 1929 wrote that his lectures to undergraduates:

were far and away the best courses on Pure Mathematics that I ever attended. He was remarkably clear, got over a great deal of ground, was interesting, and infused into his lectures a liveliness which was not characteristic of some of his contemporaries. ... In addition to his work as a lecturer he was very active in encouraging and originating research, and the great development of pure mathematics in this University is due in no small degree to the pioneering work which Glaisher began more than fifty years ago.[13]

Under Glaisher's influence both J J Thomson and the physicist James Jeans worked for a time on the theory of numbers and published their first papers in Glaisher's *Messenger of Mathematics*;[14] it has been said that Glaisher was 'the only person at Cambridge to foster Thomson's interest in research'.[15]

In introducing undergraduates to his own passion for elliptic functions Glaisher was 30 years ahead of his time in his teaching methods, for it was not until 1903 that the college created a new post, that of senior lecturer, to foster interest in mathematical research. The lectureship, which was modelled on one already instituted for classics and carried an enhanced stipend, was awarded to Alfred North Whitehead, who was required to deliver advanced lectures based on his original work.[16] Thus Glaisher was one of the first of the new breed of college lecturers who would eventually replace the traditional private coaches; *their* period of high influence was drawing to a close due to the increasingly specialised nature of the syllabuses, and the desire of the university to reform the practice of teaching.

Like his father, Glaisher was a devotee of mathematical astronomy, and consequently he was offered the post of Astronomer Royal in 1881 following the retirement of George Biddell Airy, but declined the honour; in fact he never took a permanent appointment outside Cambridge.[17] In 1883 he was appointed college tutor, and when this appointment expired in 1893 his lectureship was extended until 1901. However, by 1893 Cayley was nearing his end, and it was evident that sooner rather than later there would be a new Sadleirian Professor. As an established mathematical researcher of phenomenal output, and as an inspiring lecturer whose pupils were now making their own reputations, Glaisher had every right to be optimistic that he would be chosen to fill the vacancy when it finally arose in 1895; but he was disappointed, and the chair went to Forsyth, some 10 years his junior, who had just secured his reputation with his treatise on the functions of a complex variable.

This seems to have made Glaisher believe that, although his lectureship still had six years to run, his career was effectively over. At the age of 48 he had been institutionalised at Trinity, for as a Fellow he had a guaranteed dividend and accommodation for life, so money problems never required him to look for another post. But he foresaw that in the new century – with no tutorship or lectureship – he would have a great want of employment, and although he could have continued with his mathematical researches he was aware that he had been left behind by events, as Forsyth's triumph with complex functions confirmed. Glaisher must also have had regrets that the burden of his teaching duties was so great that he had been unable to see his books, already in proof stage, through the press; he obliquely referred to these circumstances in some of his public pronouncements. However, a few years earlier a friend had suggested that collecting pottery was an activity that he would enjoy, and so this gradually began to replace mathematics as his principal interest (see Figure 4.1).[18]

Figure 4.1 James Whitbread Lee Glaisher, surrounded by the interests of his later years.

Glaisher's academic career illustrates the stultifying effect of the Cambridge monopoly of mathematics. Forsyth commented that working with real variables only was 'the Cambridge habit',[19] yet this was far more than just a choice of working methods. The theory of functions of a complex variable, developed on the Continent in the opening decades of the nineteenth century, had become in the hands of Riemann and Weierstrass the cornerstone of modern analysis as it was then understood to be, but had been largely overlooked at Cambridge. As late as 1891 E W Hobson presented the use of complex quantities as a novel feature of his treatise on plane trigonometry, with the shade of Euler once again being invoked in his comment that:

> The subject of Analytical Trigonometry has been too frequently presented to the student in the state in which it was left by Euler, before the researches of Cauchy, Abel, Gauss, and others, had placed the use of imaginary quantities, and especially the theory of infinite series and products, where real or complex quantities are involved, on a firm scientific basis.[20]

The importance of complex analysis can be gauged from the reaction to Forsyth's book of 1893, which first made available in English the latest developments in what was by then a not-so-new branch of mathematics: Edmund Whittaker was later to declare, perhaps over-dramatically, that the book 'had a greater influence on British mathematicians than any work since Newton's *Principia*',[21] and almost certainly it secured for Forsyth the Sadleirian Chair. While the work was undoubtedly a testament to his powers of assimilation, perhaps his greatest asset was his fluency in reading French and German, a skill that Whittaker claimed was at that time almost unknown amongst mathematical lecturers, and which they showed little inclination to acquire.[22] Indeed, Forsyth's polyglotism did not stop there, for later he was to offer his publisher a book of poetry translated from Russian.[23] Complex analysis had arrived late because of Cambridge's lengthy domination of the British mathematical scene and the closed nature of the community, which together resulted in one generation of Cantabrigians teaching the next, with very little influx of new blood. There was thus a natural tendency for 'Cambridge habits' to be reinforced at the expense of the new mathematical avenues that had been explored on the Continent.

Consequently Glaisher did not make those contributions to mathematics that his early promise had suggested, although he undoubtedly encouraged and inspired his students; unlike Cayley, his prodigious output of published papers was perhaps due less to their intrinsic merit than to his editorship or co-editorship of the two journals in which they principally appeared – *The Messenger of Mathematics* and *The Quarterly Journal of Pure and Applied Mathematics*. Hardy commented that the papers were of 'uneven quality',[24] but since Glaisher himself determined the content of the journals this was hardly surprising. Although he joined *The Messenger* only as a co-editor in 1871, over the years he gained more control and from 1888 ran the journal unaided. He did not join *The Quarterly Journal* until 1879, once again as a co-editor, and once again he eventually acquired sole control, this time in 1896. He ran both journals until his death, after which *The Messenger* ceased independent publication and was absorbed by *The Quarterly Journal*, issued from Oxford under Hardy's editorship. In addition, Glaisher provided financial subsidies for the two journals out of his own pocket, and it is estimated that he wrote over a third of the pages in *The Messenger*, and even more of those in *The Quarterly Journal*; along with the *Proceedings* of the LMS, these constituted England's specialist mathematical publications. However, Glaisher's authorship did not extend only to papers. He almost completed two large books, on elliptic functions and zeta functions, and many of the pages were printed, but the duties of the tutorship that he assumed in 1883 prevented him from making the final corrections, and when his tutorship expired 10 years later 'the subject had advanced so much that much of his work was superseded'.[25] Unhappily for Glaisher, the books never saw the light of day.

Glaisher's two journals had different purposes. *The Messenger* had been founded in 1862 as *The Oxford, Cambridge, and Dublin Messenger of Mathematics*; as its name suggests, it was intended to foster communication between mathematicians in Oxford, Cambridge and Trinity College (Dublin), particularly at the graduate level, and therefore had for some years a panel of editors drawn from all three institutions.[26] In 1871 it was renamed *The Messenger of Mathematics* to attract a wider readership, and thereafter Glaisher's gradual assumption of total control reflects the dominance that Cambridge had achieved over the mathematical scene. *The Messenger* was always what Hardy called a 'minor journal'[27] in which papers from new wranglers, or even students in their final year, could comfortably be accommodated. Under Glaisher's editorship a young mathematician of promise was usually able to find an outlet: Glaisher told Hardy that he 'never discouraged a young man from rushing into print', and accordingly the first papers of Henry Baker, E W Barnes, William Burnside, E B Elliott, Forsyth, J J Thomson, James Jeans and Hardy himself all appeared there.[28] Although *The Messenger* was a 'minor journal', this did not mean that the contributions therein were of little value, for many of them were from established mathematicians of repute, and in the new century Hardy made much use of it for publishing a long series of 'Notes on Some Points in the Integral Calculus'; rather, Hardy's characterisation referred to the degree of mathematical sophistication expected of its readers. Hardy's colleague J E Littlewood, with whom he co-authored more than 100 papers, used *The Messenger* hardly at all, for he was less attracted to the art of communicating with the relatively unlearned.

That Glaisher was prepared to publish so much of his own work – about a quarter – in *The Messenger* suggests that, like many editors, he was not always able to find a sufficient quantity of new material to fill the journal: J J Thomson said that this led Glaisher 'to abandon his habit of keeping a paper back in the hope of improving it',[29] which doubtless accounted for the uneven quality that Hardy noted. The same was true of *The Quarterly Journal*, which seems to have absorbed more than half of Glaisher's output, for, as Forsyth commented, 'in these journals he could publish his own investigations as he pleased'.[30] *The Quarterly Journal* began in 1837 as the *Cambridge Mathematical Journal*, which had the double aim of encouraging research by publishing original papers, and of keeping readers up to date by abstracting the most important developments overseas.[31] After a change of name to include Dublin it ceased publication in 1854, but was resurrected the next year by Sylvester as *The Quarterly Journal of Pure and Applied Mathematics*, a title that alluded to the *Journal de Mathématiques Pures et Appliquées*, founded in 1836 by Joseph Liouville, one of the foremost French mathematicians of his day. The name of Liouville's journal was in turn borrowed from that founded by the noted German mathematician August Crelle in 1826: *Journal für die reine und angewandte Mathematik* (Journal of Pure and Applied Mathematics). Despite Crelle's intentions, however, in mid century his journal became overwhelmed by pure mathematics, and according to Felix Klein was nicknamed 'the Journal of Pure, Unapplied

Mathematics'.[32] By the time that Glaisher (working with Cayley and Norman Ferrers) had become a co-editor of *The Quarterly Journal* a similar fate had befallen it, and along with the *Proceedings* of the LMS it was recognised as a principal outlet for British pure mathematicians.

The Quarterly Journal and the *Proceedings* were about equally popular with authors, and both had considerably more pure-mathematical content than *The Messenger*. The *Proceedings*, however, was a refereed journal, and although the referees were usually drawn from the members, which might have restricted the pool of expertise available, this was never a practical difficulty because the membership included most senior British mathematicians. Glaisher, on the other hand, accepted or rejected contributions largely on his own responsibility, which caused Hardy to comment that 'the task becomes steadily more difficult as mathematics grows and specialises, and a man cannot always be worrying his friends for opinions. The position of an editor, without a properly constituted body of experts behind him, grows more and more thankless, and it is hardly likely that the experiment will be repeated.'[33] Hardy was not just being solicitous of Glaisher, for there was the implication that the difficulty of the task became increasingly evident in the quality of the journals.

Glaisher therefore occupied an increasingly powerful position with regard to the presentation of British mathematics to the world in specialist journals. He held the key that would allow a young mathematician to take the first steps in mathematical authorship, and an established author who wished to publish in a British mathematical journal other than the *Proceedings* of the LMS was almost obliged to go through him. It was perhaps detrimental to British mathematics, and cannot have been beneficial, that one whose own work was thought to have suffered through an inadequate commitment to the most recent advances in the field should have held two such influential editorships for so long. Of course, this comment refers only to the content of the journals; Glaisher also provided financial support, without which they might not have survived at all.

However, there was yet another aspect to Glaisher's involvement with mathematics, and that was through his work for the committees and councils of some of the scientific societies to which he belonged. He joined the Council of the LMS in 1872 and continued in that role until 1906, being president from 1884 to 1886. Like his father, he also served the British Association for the Advancement of Science, being for several years secretary of Section A (mathematics and physics), and president of that section in 1890.[34] In the year of his graduation he joined the Association's new Committee on Mathematical Tables, and it is a mark of the esteem in which he was already held that the other members were very eminent: Cayley, George Stokes, William Thomson (later Lord Kelvin), and Henry Smith; and in 1875 Glaisher was joined on the committee by his father.[35] Nearly always he had the preparation of some new table in hand, and this work, his principal contribution to public science, was recognised in the award of the De Morgan medal by the LMS in 1908, the Sylvester medal by the Royal Society in 1913, and honorary degrees from Trinity College (Dublin) and the Victoria University.[36] He was also active in the Royal Society and served three turns on its Council, during the last of which he was vice-president. But, as with Ranyard, the society for which he had the greatest affection was the Astronomical; he was on the council for more than 50 years, and Forsyth noted that 'his personal charm was such that for thirty-three years he was president of the Royal Astronomical Society [dining] club'.[37] Therefore his views as to the nature and purpose of pure mathematics were disseminated widely through great personal acquaintance and – for a pure mathematician – a high public profile that saw him frequently in the company of the most eminent practitioners in disciplines other than mathematics. One can get an understanding of these views from his public utterances on the subject, of which the most important are two presidential addresses: one to the LMS in 1886, and the other to Section A of the British Association in 1890.

Although Glaisher was a Cambridge man through and through, he did not think that Cambridge was conducive to the kind of research that he thought necessary for progress in pure mathematics. For this purpose he much preferred the Continental practice in which a mathematician gathered around him disciples who could learn from their master his own particular techniques: how to identify a research topic that would be susceptible to the particular expertise possessed, how to bring to bear on it not only one's own skills but also those of one's co-workers, and how to present the results in such a way as to constitute a 'proof' in the eyes of other mathematicians. Glaisher's particular concern before 1886 was the Cambridge examination system, so much so that in that year he made a very detailed history of the Mathematical Tripos examination the subject of his presidential address to the LMS, whereas most presidents chose a mathematical topic, or reflected on wider mathematical issues; however, Glaisher knew that his address would be of interest to the members because Cambridge graduates were so dominant in the Society, and the influence of the university was so great. In his opinion the system gave the student no incentive to study any topics other than those required by the syllabus: 'Think of the school of arithmeticians founded by Gauss at Göttingen, and how impossible such a result would have been at Cambridge, dominated as she has been by the competition for places in the Tripos!' But he hoped that the situation would be remedied by new regulations, which allowed a candidate to nominate a particular topic on which he wished to be examined for Part II of the Tripos: 'It will now be possible for any capable mathematician, by means of his lectures, to gather pupils round him who will bring his subject into prominence, and make it one of special study in the university.'[38] It is not clear whether Glaisher saw himself in such a role, duly armed with elliptic functions, but four years later, when he addressed the British Association, he was far less sanguine:

> Although it may not be possible to contemplate the actual position of pure mathematics in this country with any great amount of enthusiasm, I may yet feel some satisfaction in reflecting that there is more cause for congratulation at present than there has been at any time in the last hundred and fifty years, and that we are far removed from the state of affairs which existed before the days of Cayley and Sylvester. Unfortunately, we cannot point with pride to any distinct school of the pure sciences corresponding to the Cambridge school of mathematical physics, and I am afraid that the old saying that we have generals without armies is as true as ever. For this there is no immediate remedy; a school must grow up gradually of itself, as the study of mathematical physics has grown up at Cambridge.[39]

It is hard not to conclude that his own failure to attract around himself a group of disciples contributed to his disillusion with his career.

Glaisher's address to the British Association in 1890 helps us to understand the prevailing views as to the value of pure mathematics as an activity. A particular question that he addressed is why anyone should be encouraged to practise pure mathematics, and he vacillated: it was an activity that he loved and wanted to champion, but he was honest enough to recognise that the traditional justification – its support of the applied sciences – was no longer tenable. Instead, he presented pure mathematics as an independent discipline that stood or fell on its own merits, but he was unable to support this presentation with any form of argument. For example, in one of the many Senate House debates at Cambridge on the future of the Mathematical Tripos:

> [he] was surprised that Dr. Besant should suggest that Pure Mathematics should only be included [in the syllabus] on account of its applications to physical subjects. He had hoped that such a view had passed away, and that Pure Mathematics were no longer regarded as merely weapons and tools for dealing with physical questions. He maintained that the study of magnitude and numbers was as worthy of a place in University studies as the study of any physical subject. Cambridge occupied a commanding position in physical subjects, but it was unfortunately not so in Pure Mathematics.[40]

He also believed that the applied scientist was engaged in a more worthwhile enterprise, saying in his address to the British Association that 'I certainly should not wish, even if it were possible, to obtain more recruits for the pure sciences at the expense of the applied, nor do I desire to see the system of instruction which has found favour in this country so modified that pure mathematics could be carried on by narrow specialists.'[41] So he appeared to have somewhat contradictory views: although he had praised the Continental system in which a pure mathematician and his acolytes formed a cohesive 'school', with a distinctive approach to the identification and solution of research questions, he also supported a more generalist teaching tradition.

All his strands of thought concerning pure mathematics came together in this British Association address, in which the extent of his negative feelings, his personal distress at his failure to keep up with events, and his willingness to voice the arguments against his own discipline, were perfectly evident. His empathy with those who see utility as the ultimate justification for a mathematician's labours clearly struggles with his equally deep desire to study pure mathematics for its own sake:

> Passing now to the consideration of pure mathematics itself, that is to say, of the abstract sciences which can only be conquered

and explored by mathematical methods, it is difficult not to feel somewhat appalled by the enormous development they have received in the last fifty years. The mass of investigation, as measured by the pages in Transactions and Journals, which are annually added to the literature of the subject, is so great that it is fast becoming bewildering from its mere magnitude, and the extraordinary extent to which many special lines of study have been carried. To those who believe, if any such there are, that mathematics exists for the sake of its applications to the concrete sciences, it must indeed seem that it has long since run wild, and expanded itself into a thousand useless extravagances. Even the mathematician must sometimes ask himself the question – not unfrequently put to him by his friends – 'To what is it all tending? What will be the result of it all? Will there be any end?' ... To the outsider I am afraid that the subject [pure mathematics] will continue to present much the same appearance as it does now; it will always seem to be stretching out into limitless symbolic wastes, without producing any results at all commensurate with its expansion. ... It would indeed be rash to assert that there is any branch of mathematics so abstract or so recondite that it might not at any moment find an application in some concrete subject; still, it seems to me that, if the extension of the pure sciences could only be justified by the value of their application, it is very doubtful whether a satisfactory plea for any further developments could be sustained. ... Having now endeavoured to treat the proposed question [of how to justify the study of pure mathematics] impartially, and from a cold and rational standpoint, I cannot refrain from adding that, in spite of all I have said, I believe that every mathematician must cherish in his heart the conviction that at any moment some special analysis, derived in connection with a branch of pure mathematics, may bear wonderful fruit in one of the applied sciences, giving short and complete solutions of problems which could hitherto be treated only by prolix and cumbrous methods.[42]

That last comment of Glaisher's was not, as might be thought, the result of his clinging to a dying tradition: nearly 80 years later Paul Halmos, one of the leading pure mathematicians in the second half of the twentieth century, expressed similar sentiments: 'It is, I believe, a psychological fact that even the purest of the pure among us is just a wee bit thrilled when his thoughts make a new and unexpected contact with the non-mathematical universe.'[43] However, we can see from Glaisher's address the practical, human consequences of the Cambridge domination of the mathematical scene: the tension in so much of his public utterance derives from an awareness that mathematicians such as himself were in virtue of their training and institutionalisation finding it increasingly difficult to absorb and participate in new mathematical developments.

However, the sheer quantity and variety of new mathematics, to which he referred so despairingly, was not an insuperable problem for German mathematicians such as Felix Klein and David Hilbert, for they worked within an evolving system of masters and acolytes that allowed them to develop research as a team activity, and for its own sake. The novelty of this approach for British mathematicians can be judged from comments by A G Greenhill to the LMS in 1892, when he said that:

> so far as I am aware, the only attempt at the present day to introduce collaboration into mathematical work is to be found in Prof. Klein's *Seminär* at Gottingen. In his wonderful lectures ... Prof. Klein reserves his powers for a general comprehensive view of the subject; and the detailed algebraical analytical development is handed over to his pupils to work up, at the same time stimulating their interest.[44]

Klein himself said:

> I have always regarded my students not merely as hearers or pupils, but as collaborators. I want them to take an active part in my own researches; and they are therefore particularly welcome if they bring with them special knowledge and new ideas, whether these be original with them, or derived from some other source, from the teachings of other mathematicians.[45]

The academic *seminär* for teachers and their advanced students was a necessary part of Klein's system, for that was where ideas could be discussed and papers presented. It had first arisen in philology, but in 1834 a mathematics seminar was established at Königsberg University under Carl Jacobi, and by the 1860s the practice had spread to other Prussian universities and the University of Berlin,[46] although Greenhill seems to have been unaware of this. Such an approach was quite alien to the British, who clung to a vision of the mathematician as hero, engaged in a personal struggle with his symbols and equations. Such was Sylvester's attitude in his request to Cayley: 'Pray do not send me (if you find it before me) the law of the development of $\left(d/dx\right)^r \left(d/dy\right)^s$ ... as it will place me in an awkward position in publishing any memoir, if I appear compelled to *borrow* so essential a part of the investigation.'[47] Here we have the spectacle of a mathematician who chose to extend his labours by remaining in ignorance, rather than use the findings of his friend; anything more contrary to Klein's point of view is difficult to imagine.

The heroic idea of mathematics also forced mathematicians to spend time in laborious calculations that should have been undertaken by students, leaving the mathematician himself free to address points of principle. This was a matter of concern to some Victorians such as Greenhill, who in 1892 remarked that 'Prof. Tait makes a feeling complaint somewhere that in looking back at the time spent on working out a new development the greater part of it is occupied on analysis that might well be turned over to a trained mathematical student.'[48] And in 1896 Percy MacMahon compared Cambridge practice unfavourably with that of the Germans, with an implication that even Cayley's work had been affected:

> No mathematician of Cayley's time possessed comparable powers over long expressions and tiresome calculations. He was a great intuitionist, but his reputation in this respect may have suffered from his extraordinary skill as a formalist. Much of his work might, and should, have been relegated to others. This, without doubt, would have been so had there existed a school of pure mathematics at Cambridge, similar to those in many German Universities. These remarks arise from my recollections of the vast amount of purely numerical labour, that might have been handed over to a calculator, undertaken by Cayley about the year 1885 when engaged in calculating ground forms and syzygies of binary quantics. ... With a lesser man it [the power of calculating] would have become a vice. In truth, it has been well said that attention to details is frequently the refuge of a man who is destitute of, or who has lost, the power of discussing broad and general questions.[49]

Thus under the German system Glaisher might have found more time to keep abreast of relevant modern developments, and would have had disciples with whom to discuss them. Furthermore, in Germany non-constructive methods of 'proof' were being developed, and so (in Karen Hunger Parshall's phrase) mathematicians there were able 'to skim just above the level of explicit calculation in their work';[50] whereas British mathematicians, many of whom regarded these new forms of 'proof' with suspicion, clung to the heroic methods of constructive proof that had been so productive for Cayley and Sylvester.[b]

Therefore Glaisher was a transitional figure. On the one hand, being the son of his father, he was well aware of the merits of the natural philosopher, and accordingly abhorred moves towards a narrow specialisation in pure mathematics that might limit the reservoir of talent available to the applied sciences. On the other hand, his own talents and inclinations led him to pursue, and encourage his students to undertake, pure mathematical research at a time when this was a novelty at Cambridge, and his conception of the importance of this led him to support the reform of the Mathematical Tripos and the abolition of the order of merit. But his research options were limited by the methods that he had learned at Cambridge, which meant that his treatment of elliptic functions, a topic of great contemporary interest, did not have the impact that a more modern approach would have produced under the aegis of complex analysis. He was firmly committed to the belief that pure mathematics was an activity worth pursuing for its own sake, but lacked the confidence that this was intellectually respectable position; and he was unable to comprehend a view that did not take the application of some part of pure mathematics as the greatest reward that could be desired. Consequently, in public he appeared to vacillate, because his honesty forced him to recognise and draw attention to the insufficiency of traditional arguments in favour of pure mathematics, whilst the respect, almost amounting to awe, in which he held the applied sciences led him to avoid any form of words that might appear to detract from them. His support for pure mathematics therefore appeared to be a matter of personal conviction, but nothing more.

However, he was not the only Cambridge mathematician who became discouraged when unable to keep abreast of the most recent developments, and Henry Baker, who enjoyed a long life from 1866 to

---

[b] Some theorems assert the existence of a mathematical entity with certain properties. A constructive proof constructs and exhibits a mathematical entity with those properties; a non-constructive proof is an exercise in logic showing that a mathematical entity with those properties *must* exist.

1956, provides another example, although with a mathematically happier outcome. After spending 25 years working on the theory of functions, he realised that his work was unappreciated and that he was unable to attract students.[51] He had been trained in the old school, was out of sympathy with the approach to analysis being successfully promulgated in the new century by Hardy and Littlewood, and in 1910 wrote to his friend and confidant Joseph Larmor: 'I am bound to answer your letter, which is a relief to a lonely spirit discouraged ... I must not trust myself to say more; it is hard to resist the impression that I am judged to have misspent many years.' By 1913 he seemed on the verge of a mental breakdown: 'I want to live in a family, in a village, to watch how the sweet peas grow, to nod to all the folk I pass, and give 6^d to the Porters at the station, when 2^d would do ... This seems all off any rails.' But he did not try to resolve his problems, as Glaisher had done a decade earlier, by becoming less involved with mathematics. Instead he switched the focus of his research to geometry, and shortly before the First World War was appointed a controversial Lowndean Professor of astronomy and geometry – controversial, because the chair was traditionally held by an astronomer rather than a geometer. Yet the change seemed to bring him little real happiness; on Larmor's retirement in 1932 Baker wrote: 'I should most gladly go out of the turmoil (10 years your junior) tomorrow, if I could.'[52]

Once appointed, however, Baker lost little time in instituting a school of geometry that gained international recognition and still existed in 1951.[53] It was maintained by his weekly tea parties, during which the latest topics in geometrical research could be discussed; these became the nearest things to mathematical seminars that Cambridge then had, although they were somewhat different from the very formal German variety, which included the reading of papers – something that Baker wanted to avoid. Here we see the first recognition amongst Cambridge mathematicians that training in research required a forum in which staff and students could become apprised of each other's work and discuss it critically. However, Baker was not the first at Cambridge to introduce such a gathering; that award goes to John Seeley, who was appointed Regius Professor of Modern History in 1869 and invited 'advanced students to his home to participate in discussion seminars and informal conversation classes'.[54] Thus both Glaisher and Baker suffered during the first part of their careers from the stultifying effects of Cambridge pure mathematics, but Baker's professorial tea parties

129 / The Pure Mathematician as Hero

Figure 4.2 Henry John Stephen Smith, captured in an unusually informal early portrait taken c. 1849 by his friend Nevil Story Maskelyne; it shows a side to Smith's character very different from that conveyed by the grave bearded figure of later, more formal, portraits.

yielded considerable success, for amongst his doctoral students were H S M Coxeter and Daniel Pedoe – two future geometers of distinction – as well as the polymath and broadcaster Jacob Bronowski.

It is interesting to compare Glaisher with his friend Henry John Stephen Smith (see Figure 4.2), the Oxford mathematician, particularly as a most unfortunate set of circumstances during the last year of Smith's life, in which Glaisher was personally involved, illustrates the problems that can arise through a lack of international communication. Glaisher and Smith had some common mathematical interests (mainly theta functions and elliptic transformations),[55] but whereas Glaisher pursued these by traditional Cambridge methods, Smith spent his whole career at Oxford, where mathematics was of little account, and consequently he was much more free to absorb Continental developments: as Glaisher wrote, 'The subjects in which [Smith] was a master, and to which his own contributions were of such high value, were quite beyond the range of Cambridge mathematical teaching.'[56] Smith was not a natural philosopher in the Cambridge mould, and he became an enthusiastic proponent of the study of mathematics for its own sake, rather than because of its utility.[57]

There are two reasons why he was not as influential as might have been expected in establishing the role of the pure mathematician. The first is his decision to study at Oxford for a degree, which may seem surprising given its attitude towards mathematics. However, Oxford was outstanding in the classics, and it was that branch of learning in which Smith excelled as a boy. He graduated in 1849 with the highest honours in both classics and mathematics, and it was only then that he decided to specialise in the latter, saying that his choice was partly dictated by his weak eyesight, the logic being that mathematics involved more thinking and less reading;[58] nevertheless, this did not stop him from publishing two mathematical papers in Latin.[59] He remained at Oxford all his life, and as potential mathematicians naturally gravitated towards Cambridge he never gathered around him students who could have formed a 'school' or otherwise have given greater exposure to his mathematical achievements. Nor did he have colleagues of any mathematical distinction with whom he could enjoy informal discussions, and his closest confidant in such matters was Glaisher at Cambridge, who wrote after Smith's death in 1883: 'I believe I was the only person who ever knew him *really as he was* as a mathematician & to whom he opened his heart.'[60]

The second reason for Smith's lack of influence is that his publications were less numerous than might have been expected, because for much of his university career (which lasted until his death) the bulk of his time was spent in official and administrative duties, with his mathematical researches being undertaken in the hours that remained, often late at night when the day's work would otherwise have been over – and this notwithstanding that he was recognised, although more on the Continent than in Britain, as one of the foremost mathematicians of his generation. Until the final decade of his life his college was Balliol, where the dominant personality was Benjamin Jowett, and Hardy commented on 'Jowett's delightfully naive memoir of Henry Smith ... Smith, says Jowett, "was not the author of any considerable work"; he "lived and died almost unknown to the world at large." I wonder how many Germans knew the work of Smith, for one who had ever heard of Benjamin Jowett....'.[61] At Oxford, Jowett was an outstanding classical scholar, Regius Professor of Greek from 1855, Master of Balliol from 1870, and Vice-Chancellor from 1882 to 1886, but his unwillingness to appreciate the value of learning in fields other than his own inspired H C Beeching's ditty:

> First come I; my name is Jowett
> There's no knowledge but I know it.
> I am Master of this college:
> What I don't know isn't knowledge.

Consequently he was not a man to have recognised Smith's achievements in a discipline that was so totally foreign to him.

Smith was the only mathematician of international standing that Oxford produced during the Victorian era, and he made a particular study of the theory of numbers. In 1868, and following work by Eisenstein, he published in the *Proceedings of the Royal Society*, but without full proofs, formulae that determined the hitherto unsolved question as to the conditions under which an integer could be expressed as the sum of five or seven squares.[62] The French Academy was unaware of his work, and in 1881 set the five squares problem for its Grand Prix des Sciences Mathématiques, the award to be made in 1883. Smith discovered this only in 1882, and after canvassing Glaisher's opinion he wrote to the Academy, pointing out that 14 years earlier he had solved the problem and published a solution. It is not clear whether Smith read of the competition himself in the *Comptes Rendus Hebdomadaires des Séances de l'Académie des Sciences*, or whether his attention was drawn to it by someone else, but that he felt obliged to turn to his Cambridge friend for advice demonstrates the extent of his mathematical isolation at Oxford. In reply to his letter the mathematician Charles Hermite, who was responsible for the competition, asked Smith to relieve the Academy's embarrassment by submitting as an entry both the solution and the proof, after which the Academy would be apprised of his priority. Smith duly obliged, but it was not until two months after he died that the distinguished analyst Camille Jordan reported on the entries. Due to an oversight on Hermite's part Jordan was not aware of Smith's earlier work, and consequently awarded the prize jointly to Smith and Herman Minkowski, then a student at Königsberg. Jordan's report remarked that the entries were similar, and once Smith's priority was made public the Academy was widely criticised, not only because it was unaware of work published by the Royal Society in 1868, but also because of suggestions in the newspapers that Minkowski did have prior knowledge of it, which accounted for the similarity of the entries. There has never been any evidence to support this allegation.[63]

Notable in this affair is the time that elapsed between the first announcement of the competition and Smith's becoming aware of it. Certainly anyone who knew both his paper and the terms of the competition would not have left him in ignorance, since the situation had almost unlimited potential for embarrassment. However, in Britain his work on the theory of numbers was unique in the extent to which it followed developments on the Continent, a fact to which Hermite alluded in his reply to Smith's letter.[64] Thus Smith's paper, although published by the Royal Society, was of interest to very few people in Britain, whereas in terms of subject matter it would clearly have been of the greatest interest to many Continental mathematicians had they been aware of it, since the question was deemed sufficiently worthy to be the subject of the Academy's Grand Prix. On the other hand, the announcement of the Competition was in the *Comptes Rendus*,[65] which suffered, as far as the British were concerned, from being in French. Thus in this incident national barriers worked both ways: an important paper by Smith was overlooked on the Continent because it was published in Britain, and an important announcement relating to its subject matter was overlooked in Britain because it was published in France.

Smith, perhaps more than anyone else in Britain, was aware of the gulf that existed between British mathematics and that which was developing so rapidly on the Continent. He gave unrestrained expression to his views in a letter to Isaac Todhunter:

> All that we have, one may say, comes to us from Cambridge; for Dublin has not of late quite kept up the promise she once gave. Further, I do not think that we have anything to blush for in a comparison with France; but France is at the lowest ebb, is conscious that she is so, and is making great efforts to recover her lost place in Science. ...
>
> But in Pure Mathematics, I must say that I think we are beaten out of sight by Germany; and I have always felt that the *Quarterly Journal* is a miserable spectacle, as compared with Crelle, or even

Clebsch and Neumann [the journal *Mathematische Annalen*]. Cayley and Sylvester have had the lion's share of the modern Algebra (but even in Algebra the whole of the modern theory of equations, substitutions, &c., is French and German). But what has England done in Pure Geometry, in the Theory of Numbers, in the Integral Calculus? What a trifle the symbolic methods, which have been developed in England, are compared with such work as that of Riemann and Weierstrass!

Smith was very much alive to the significance of these two German mathematicians, who were largely responsible for the rigorous formulation of analysis and the calculus with which we are familiar today. However, their work was almost unknown amongst British mathematicians generally. Smith continued:

But it is with the younger, or at least the less-known, people that I feel the difference most. Our English papers are so often quite free from anything really new, whereas a German takes care to know what is known before he begins to work, and besides generally takes care to work at some really important problem, and not at some trifling expression for the co-ordinates of the focus.

If I had room, I should vent my spleen (or perhaps my envy) by saying that I attribute the mischief to the business of problem-making; ninety per cent. of the good problems in Pure Mathematics that I see, are, if I mistake not, mere fragments of some great theory, of which the candidate is supposed to be ignorant.[66]

A comparison of the careers of the two friends Glaisher and Smith therefore leads to a paradoxical result. Glaisher, although the younger man, found himself institutionalised in Britain's mathematical powerhouse and enslaved by its traditional modes of thought. Consequently, burdened by teaching duties, he found himself unable to keep abreast of burgeoning developments on the Continent, and eventually collecting pottery displaced mathematics as his principal interest. Smith, who took a first in both mathematics and classics, was similarly burdened, but had not been trained in the same fashion and so was able to absorb and stay in tune with Continental mathematical advances to an extent that was unmatched in Britain. Although his output was far less than Glaisher's, it was far superior in originality, and it is a fair assessment of the stultifying power of Cambridge to say that that Smith would not have been able to flourish there as he did in the mathematically insignificant confines of Balliol College, Oxford.

If Henry Smith's mathematical achievements benefitted from his mental and physical isolation from Cambridge, those of the soldier Percy MacMahon were only made possible by a Cambridge connection, although his life was spent mostly outside academia, and so was a complete contrast to the lives of Glaisher, Baker and Smith. He was the son of a Colonel, grew up in military surroundings, and seems to have been destined for a military career from an early age. He was a promising mathematician at school, and in 1871, at the age of 16, began studies at the Royal Military Academy (RMA), Woolwich, as a gentleman cadet. Mathematics had always been the discipline carrying the most marks there on account of its many applications in ballistics and other matters of concern to the military,[67] but in the words of an 'Old Cadet': 'The various subjects inspired about the same sentiments as they do now. Mathematics were regarded with awe and some distaste, as necessitating thought.'[68] Something of the flavour of the mathematical education that MacMahon enjoyed can be gathered from the following reminiscence of Sylvester's class. Although MacMahon would not have witnessed these particular scenes because he did not join the Academy until two years after Sylvester had been forced on age grounds to retire, the anecdotes show the general standard of the class and of the seriousness with which mathematical instruction was regarded. At this time Sylvester was, after Cayley, probably Britain's foremost pure mathematician.

> We began [when we rejoined in 1864] surveying and higher mathematics, the latter under Professor Sylvester, a splendid mathematician, but totally incapable of teaching cadets. As a natural result, order was usually badly kept in his Academy, and sundry measures of annoying him were indulged in with success by the cadets. One plan which was occasionally tried, was for a large number of them to drop down behind their desks. Sylvester would suddenly awake from the solution of some abstruse problem and see the class-room half empty. This made him rush up and down, a movement which was prepared for by sprinkling the floor

round his table with wax matches, which went off in succession as he stamped round, driving him quite wild. Another trick was to fill his ink-bottle with chalk, which clogged his pens and made him mad! But with all his little ways, he could teach well if he was allowed his own method, and personally I owe a good deal to him.[69]

That Sylvester was prepared to put up with such a routine for 14 years suggests that teaching opportunities were not easily come by, although an additional inducement was that the RMA, as an arm of government, offered attractive pay when compared with some of the more exalted alternatives: in 1856 the professor of mathematics received an annual salary of £500.[70]

After two years MacMahon had completed the syllabus, and in March 1873 was posted to India.[71] Four years later he came home on sick leave that lasted 18 months, and after some further service in England returned to Woolwich in 1880 to join the advanced class at the Department of Artillery Studies. This class dated from 1864, when changes in artillery practice required the establishment of 'a special course of instruction, lasting two years and entered by means of a competitive examination'. The course of study always comprised 'Mathematics, Mechanism, Chemistry, Metallurgy, and Physics as applied to the matériel of the day together with a thorough investigation of that matériel.' However:

> Mathematics has always predominated, not only on account of its application to ballistics, but also because the natural and physical sciences in their progress make ever increasing demands on its aid. Thus even when the Class was first instituted the lowest standard that could be usefully adopted included the calculus and its elementary applications to statics and dynamics. This, however, was a standard beyond the reach of many of the students who had to spend time in acquiring mathematical knowledge rather than in its application, so that as a whole the Classes could not be said to be carrying out the purpose for which they had been designed.[72]

Consequently there was much subsequent tinkering with the arrangements. MacMahon, promoted to captain in 1881, passed the examination in 1882, but the standard reached, and the subjects studied, were clearly not such as to immerse him in those branches of mathematics with which he was later to make his reputation. Nevertheless his competence was immediately recognised, as he was forthwith appointed an instructor in mathematics, a post that he was to retain for six years; and 1882 also saw the publication of his first mathematical paper.

It was then that MacMahon first met invariant theory, which was one of the main concerns of British pure mathematicians, and to which he was to devote much of his intellectual energy for the next 20 years. Decades later Joseph Larmor, in a letter to *The Times*, said that Alfred Greenhill, who was professor of mathematics in the department of artillery studies from 1876 to 1907, introduced MacMahon to these ideas.[73] Greenhill was a graduate of St John's, Cambridge, having been second wrangler and joint Smith's prizeman in 1870, and so for more than 30 years provided at a senior level a point of contact between the Academy and Cambridge mathematics. Before his appointment he had been professor of applied mathematics at Cooper's Hill Engineering College and gained a knighthood for services to gunnery; he was 'a Newtonian for whom the *Principia* remained unrivalled'.[74] Yet in spite of this lifelong devotion to applied mathematics, he recognised that MacMahon had mental capabilities that were ideally suited to the requirements of British invariant theory. Larmor, in his letter, wrote of MacMahon that:

> The young captain threw himself with indomitable zeal and insight into the great problems of this rising edifice of science; and in a very short time he was to be counted as conspicuous amongst the leaders, largely by the invention of new methods of approach. So complete was his scientific absorption then, and during successive tenures as Instructor and Professor at Woolwich, that one was accustomed to hear his military friends refer in chaff to him as 'a good soldier spoiled.'[75]

Certainly such studies formed part of the syllabus at neither the RMA nor the department: not only did invariant theory have no military application, but also it was way beyond the understanding of those who were being taught, for the students' level of mathematical competence was not high. However, one personal trait may have made the study of invariant theory particularly pleasing to MacMahon, and that was his ability to undertake complex mental calculations quickly and accurately. This is a highly desirable attribute for an invariant theorist, and his ability as a calculator was reported as being even greater than that of the Indian mathematician Srinivasa Ramanujan, who was himself famous for doing most of the 'workings' for his results in his head. This intelligence has come via a rather indirect route, as Gian-Carlo Rota records:

> With his moustache, his 'British Empah' demeanor, and worst of all his military background, MacMahon was hardly the type to be chosen by Central Casting for the role of the Great Mathematician. Ramanujan, with his Eastern aura, his frail physique, and his swarthy good looks qualified all the way. It would have been fascinating to be present at one of the battles of arithmetical wits at Trinity College, when MacMahon would regularly trounce Ramanujan by the display of superior ability for fast mental computation (as reported by D. C. Spencer, who heard it from G. H. Hardy). The written accounts of the lives of the characters, however, omit any mention of this episode, since it clashes against our prejudices.[76]

In 1888 MacMahon left his post as instructor to become inspector of Warlike Stores at Woolwich, and in 1889 was promoted to major. This new position lasted only until 1891, when he was appointed instructor (later professor) of electricity at Artillery College (as the department had been renamed), where he remained until he retired from the army in 1898. From 1906 to 1920 he acted as deputy warden of standards at the Board of Trade, and in addition Larmor tells us that 'On being co-opted by the University of Cambridge to the honorary degree of Doctor of Science in 1904 [MacMahon] attached himself by invitation to St. John's College, where he had acquired many friends.'[77] This attachment lasted until the 1920s, when ill health obliged him to remove to Bognor Regis, where he died on Christmas Day 1929.

MacMahon remained mathematically productive until his closing years, and not unnaturally the period after his retirement from the army provided the greatest output of papers. In the eyes of his contemporaries he was the British mathematician upon whom naturally fell the mantle of Cayley and Sylvester, the foremost British proponents of invariant theory, after their demise in the 1890s. However, by then Continental mathematicians such as Paul Gordan and David Hilbert, using much more general methods than their British counterparts, had overcome most of the challenges that the theory presented, although later some of its underlying ideas were to return with a new guise in the theory of groups and other algebraic constructs. Accordingly, in the opening years of the twentieth century the notation, vocabulary and methods that had been characteristic of British invariant theorists were of little interest, and the mathematical vacuum was filled by the analysis of Hardy and Littlewood. Fortunately for MacMahon, his interests had by then expanded beyond invariant theory to encompass the theory of permutations and combinations, and late in life he developed an interest in mathematical games and puzzles, which led to a modest book describing amusements of his own devising on the subject of repeating patterns.[78] However, he is now best remembered for a two-volume treatise on combinatory analysis that the American Mathematical Society still keeps in print.[79] This branch of mathematics, like invariant theory, requires a very high level of computational skill, and the story of his contests with Ramanujan suggests that such skill was his greatest asset, and that he was drawn to those branches of mathematics that gave him the best chance of exercising it; in him there was an entirely fortuitous conjunction of a particular cast of mind and appropriate subject matter, discovered in a serendipitous way through Greenhill.

The Cambridge mathematical tradition, whether operating directly and pervasively as in the cases of Glaisher and Baker, or indirectly as with MacMahon, offered a world that had distanced itself from new approaches to mathematics that were being explored in Germany, France and Italy. As we saw in the case of Henry Smith and the Grand Prix, the loss was not all on one side; nevertheless, it was the British who lost the most, because they were not participating in significant developments concerning the nature of mathematical proof, the handling of the infinite, and the logical foundations of mathematics. From these debates the British had effectively debarred themselves, and it was not until the new century that they were able to make any significant contributions.

There was also a language barrier at work: English was not the primary means of communication for most mathematical authors worldwide, and 120 of the 142 journals surveyed by the *Jahrbuch über die Fortschritte der Mathematik* in 1890 were in some other language. Unfortunately, British mathematicians were no more inclined than the British generally to learn foreign languages, and Edmund

Whittaker, second wrangler in 1895, remarked that 'they could not read German and would not read French'.[80] Although obviously an exaggeration his comment nevertheless contained a good deal of truth; in the normal course of events the majority of British mathematicians would not have found themselves perusing any of those 120 journals that contained mathematical papers in a foreign language, and so British mathematical traditions were eroded less than they would otherwise have been. The excitement caused by Forsyth's 1893 treatise on complex functions arose because for the first time it made available in English a comprehensive survey of one of the most important branches of pure mathematics.

One consequence was that few British authors published pure mathematics in foreign-language journals, with an average annual total output of less than 25 pages. Likewise, with most of the important mathematical journals being published in other languages, there was little reason for a Continental pure mathematician to publish in English, a circumstance reflected by the content of the *Proceedings* of the LMS. Thus the tendency was for British pure mathematicians to publish in British journals that would be read mainly by other British mathematicians, while those journals contained little by way of leavening from Continental mathematicians. As a result British pure mathematicians were accustomed to obtaining their inspiration mainly from other British mathematicians; but equally, Continental pure mathematicians were not necessarily alive to the work undertaken by their confreres in Britain.

Glaisher had recognised that the Continental system of master and acolytes was far more conducive to mathematical research than the individualistic Cambridge tradition, which bred 'generals without armies'. However, he failed to realise that the Continental system would have enabled the British to cope not only with the proliferation of new ideas in pure mathematics to which he had reacted so despairingly, but also with the demands of time that the routine labour of calculation placed on mathematicians who worked in a culture that prized explicit results and was suspicious of existence proofs. He simply trusted that improved means of communication between British and Continental workers would develop so as to expedite research, but offered no grounds for this trust; yet the signs were there that the days of the traditional British approach were numbered. At Oxford, Sylvester – who had returned to England from America in 1884 to succeed Smith in the Savilian chair – was a founder of the Oxford

Mathematical Society, which had a rule that the papers read should *not* be printed; the aim was to concentrate on the discussion of research topics and the interchange of ideas.[81] Sylvester perhaps understood what was happening at the LMS, where an undue concentration on reading long and very technical papers had by the 1890s virtually nullified any social function that the Society's meetings may originally have possessed. On the other hand, Felix Klein in Germany had developed organised research for the production of new mathematical results: in his seminar he laid down the programme, planned the work and provided overall guidance and control, but made individual researchers responsible for undertaking particular tasks. Thus he was freed from the routine business of detailed computation that took up much of the time of British pure mathematicians, and did not himself have to master the intricacies of every new piece of research. However, a necessary consequence was that a mathematician had to be prepared to give every credit to his co-workers, and could no longer claim to be the only begetter of his paper.

Consequently, throughout the second half of the Victorian era, Cambridge continued to dominate the British mathematical scene. With little influence from elsewhere, British pure mathematicians tended to use traditional Cambridge methods that were not at the forefront of modern research, as the careers of Glaisher, Baker and MacMahon demonstrated. Consequently, British mathematical journals failed to attract the patronage of Continental mathematicians, which in turn just increased British insularity; and for this vicious circle to be broken, it was necessary to absorb new concepts and ways of working that had been developed elsewhere. This meant that British pure mathematicians needed to understand their discipline as being concerned with logical development rather than empirical truth, and as using a methodology that embraced not only indirect or non-constructive proofs but also much more compact symbolic representations. It was all very different from the explicit constructive demonstrations by a lone mathematician that had for so long been the hallmark of British mathematics, and in the new century the scene in Britain was to be dominated by two famous partnerships: those of Whitehead and Russell in mathematical logic, and of Hardy and Littlewood in analysis. In pure mathematics, the end of the century and the end of the reign brought also an end to the Victorian veneration of the individual mathematical achiever – the pure mathematician as hero.

Notes and References

1. A R Forsyth, 'James Whitbread Lee Glaisher', *Journal of the London Mathematical Society*, 4, 2 (1929), 101–112, at 109.
2. See the abstract of Glaisher's will in *The Cambridge Review*, L, 1240 (24 May 1929), 470.
3. *Oxford Dictionary of National Biography*, Art. 'Glaisher, James'.
4. J W L Glaisher, 'Tables of the Numerical Values of the Sine-Integral, Cosine-Integral, and Exponential Integral', *Philosophical Transactions of the Royal Society of London*, 160 (1870), 367–388; Forsyth, 'James Whitbread Lee Glaisher', 103.
5. Andrew Warwick, *Masters of Theory: Cambridge and the Rise of Mathematical Physics* (Chicago: University of Chicago Press, 2003), p. 503.
6. Forsyth, 'James Whitbread Lee Glaisher', 110.
7. Forsyth, 'James Whitbread Lee Glaisher', 101, 110.
8. Leonard Roth, 'Old Cambridge Days', *American Mathematical Monthly*, 78 (1971), 223–236, at 231.
9. G H Hardy, 'Dr. Glaisher and the "Messenger of Mathematics"', *Messenger of Mathematics*, LVIII (1929), 159–160, at 160.
10. Forsyth, 'James Whitbread Lee Glaisher', 111.
11. J W L Glaisher, 'The Mathematical Tripos', *Proceedings of the London Mathematical Society*, XVIII (1886), 4–38, at 4–5.
12. In a passage written by Forsyth for H H Turner, 'James Whitbread Lee Glaisher', *Monthly Notices of the Royal Astronomical Society*, LXXXIX, 4 (1929), 300–308, at 302–303. The obituary is signed 'H H T'; for the attribution to Turner, see Glaisher's entry in the *Oxford Dictionary of National Biography*.
13. J J Thomson, 'Dr. Glaisher', *The Cambridge Review*, L, 1228 (1929), 212–231, at 212. The obituary is signed 'J J T'.
14. Hardy, 'Dr. Glaisher', 160; also George Paget Thomson, *J J Thomson and the Cavendish Laboratory in his Day* (London: Nelson, 1964), p. 22.
15. E A Davis and I J Falconer, *J J Thomson and the Discovery of the Electron* (London: Taylor & Francis, 1997), p. 9.
16. Victor Lowe, *Alfred North Whitehead: The Man and his Work*, 1 (Baltimore, MD: Johns Hopkins, 1985), p. 212.
17. A R Forsyth, 'Glaisher, James Whitbread Lee', in J R H Weaver (ed.), *The Dictionary of National Biography 1922–1930* (Oxford: Oxford University Press, 1937), p. 339.
18. Julia E Poole, 'Dr J W L Glaisher: The Making of a Great Collection', in Michael Archer, *Delftware in the Fitzwilliam Museum* (London: Philip Wilson, 2013), p. xvi.
19. Forsyth, 'James Whitbread Lee Glaisher', 110.
20. E W Hobson, *A Treatise on Plane Trigonometry* (Cambridge: Cambridge University Press, 1891), pp. vi–vii.
21. E T Whittaker, 'Andrew Russell Forsyth', *Obituary Notices of Fellows of the Royal Society*, 4 (November 1942), 208–227, at 218.
22. Whittaker, 'Forsyth', 217.
23. Letter from Forsyth to Macmillan & Co, 8 December 1925: Macmillan Archives (British Library), Add 55197, fo. 60.
24. Hardy, 'Dr. Glaisher', 160.
25. Thomson, 'Dr Glaisher', 212.
26. Sloan Evans Despeaux, 'Launching Mathematical Research without a Formal Mandate: The Role of University-Affiliated Journals in Britain', *Historia Mathematica*, 34 (2007), 89–106, at 100–102.

27. Hardy, 'Dr. Glaisher', 160.
28. Hardy, 'Dr. Glaisher', 159–160.
29. Thomson, 'Dr Glaisher', 212.
30. Forsyth, 'James Whitbread Lee Glaisher', 111.
31. Duncan Gregory, 'Preface', *Cambridge Mathematical Journal*, I (1839), 1. The first volume of the *CMJ* comprised issues spanning two years, and Gregory's Preface, originally anonymous, appeared in the first issue (November 1837). See Sloan Evans Despeaux, '"Very Full of Symbols": Duncan F. Gregory, the Calculus of Operations, and the *Cambridge Mathematical Journal*', in Jeremy Gray and Karen Hunger Parshall (eds.), *Episodes in the History of Modern* Algebra *(1800–1950)* (Providence, RI: American Mathematical Society and The London Mathematical Society, 2007), pp. 51, 70.
32. Felix Klein (tr. M. Ackerman), *Development of Mathematics in the 19th Century* (Brookline, MA: Math Sci Press, 1979), p. 86; the original version was published in 1928.
33. Hardy, 'Dr. Glaisher', 159.
34. Forsyth, 'Glaisher, James Whitbread Lee', 339.
35. Sidney Lee (ed.), *The Dictionary of National Biography 1901–1911* (London: Smith Elder, 1912), Art. 'Glaisher, James'.
36. Thomson, 'Dr. Glaisher', 212.
37. Forsyth, 'Glaisher, James Whitbread Lee', 339.
38. Glaisher, 'The Mathematical Tripos', 32, 33.
39. J W L Glaisher, 'Presidential Address to Section A', in *Report of the Sixtieth Meeting of the British Association for the Advancement of Science* [1890] (London: John Murray, 1891), p. 724.
40. Discussion in the Senate House on 12 March 1890: *Cambridge University Reporter*, 18 March 1890, 559.
41. Glaisher, 'Presidential Address to Section A', 724–725.
42. Glaisher, 'Presidential Address to Section A', 721–723.
43. P R Halmos, 'Mathematics as a Creative Art', *American Scientist*, 56, 4 (1968), 384.
44. A G Greenhill, 'Collaboration in Mathematics', *Proceedings of the London Mathematical Society*, XXIV (1892), 5–16, at 13.
45. Felix Klein, *The Evanston Colloquium: Lectures on Mathematics delivered from Aug. 28 to Sept. 9, 1893 before Members of the Congress of Mathematics, held in connection with the World's Fair in Chicago, at Northwestern University, Evanston, Ill* (New York: Macmillan, 1894), p. 96.
46. Gert Schubring, 'Germany to 1933', in I Grattan-Guinness (ed.), *Companion Encyclopedia of the History and Philosophy of the Mathematical Sciences*, 2 (London: Routledge, 1994), pp. 1447, 1443.
47. Letter from Sylvester to Cayley dated 2 December 1854, in Karen Hunger Parshall, *James Joseph Sylvester: Life and Work in Letters* (Oxford: Clarendon Press, 1998), p. 76.
48. Greenhill, 'Collaboration', 12.
49. P A MacMahon, 'Combinatory Analysis: A Review of the Present State of Knowledge', *Proceedings of the London Mathematical Society*, XXVIII (1896), p. 7.
50. Karen Hunger Parshall, 'The One-Hundredth Anniversary of the Death of Invariant Theory?', *Mathematical Intelligencer*, 12, 4 (1990), 12.
51. W D V Hodge, 'Henry Frederick Baker (1866–1956)', *Biographical Memoirs of Fellows of the Royal Society* 2 (1956), 49–68, at 56.
52. Letters from Baker to Larmor dated 28 February 1910, 14 July 1913 and 14 June 1932: Library of St. John's College, Cambridge, Miscellaneous Papers BA4.

53. H R Hassé, 'My Fifty Years of Mathematics', *The Mathematical Gazette*, XXXV, 313 (September 1951), 161.
54. Sheldon Rothblatt, *The Revolution of the Dons: Cambridge and Society in Victorian England* (London: Faber, 1968), p, 180; R T Shannon, 'Seeley, Sir John Robert', in *Oxford Dictionary of National Biography* 2004 (available online at: https://doi.org /10.1093/ref:odnb/25025).
55. See Glaisher's introduction to J W L Glaisher (ed.), *The Collected Mathematical Papers of Henry John Stephen Smith M.A., F.R.S., late Savilian Professor of Geometry in the University of Oxford*, I (Oxford: Clarendon Press, 1894), p. lxxxvi.
56. J W L Glaisher, 'Notes on the Early History of the Society', *Journal of the London Mathematical Society*, 1, 1 (1926), 51–64, at 54.
57. For a sketch of Smith's character as a young man, and the origin of the portrait reproduced in this book, see Vanda Morton, *Oxford Rebels: The Life and Friends of Nevil Story Maskelyne 1823–1911, pioneer Oxford Scientist, Photographer and Politician* (Gloucester: Alan Sutton, 1987), pp. 94–96.
58. See Pearson's Biographical Sketch in Smith's *Collected Mathematical Papers*, I, p. xviii.
59. 'De Compositione Numerorum Primorum Formae 4λ+1 Ex Duobus Quadratis', *Crelle's Journal*, L (1855), 91–92; 'De Fractionibus Quibusdam Continuis', in *Collectanea Mathematica (in memoriam Domenici Chelini)*, (Mediolani: Ulrici Hoepli, 1881), pp. 117–143.
60. Letter from Glaisher to Thomas H Escott, 17 April 1883: Escott Papers (British Library), Add. 58780, fo. 55.
61. G H Hardy, 'Mathematics', *The Oxford Magazine*, 48 (1930), 819; Jowett's memoir is in Smith's *Collected Mathematical Papers*, I, pp. xxxvii–xlv.
62. H J S Smith, 'On the Orders and Genera of Quadratic Forms containing more than three Indeterminates', *Proceedings of the Royal Society*, XVI (1867), 197–208.
63. A full account of the whole episode, together with copies of relevant correspondence, is given by Glaisher in his introduction to Smith's *Collected Mathematical Papers*, I, pp. lxv–lxxi.
64. Hermite wrote to Smith: 'Vous êtes seul en Angleterre à marcher dans la voie ouverte par Eisenstein. M. Kronecker est seul en Allemagne ; et chez nous M. Poincaré qui a jeté en avant quelques idées heureuses sur ce qu'il appelle les invariants arithmétiques, semble maintenant ne plus songer qu'aux fonctions Fuchsiennes et aux équations différentielles.' Letter dated 26 February 1882, in Smith's *Collected Mathematical Papers*, pp. lxvi–lxvii.
65. *Comptes Rendus Hebdomadaires des Séances de l'Académie des Sciences*, XCII, 622, and XCIV, 330. It was the latter announcement that led to Smith's letter.
66. Letter from Smith to Todhunter quoted by Glaisher (no date given) in his introduction to Smith's *Collected Mathematical Papers*, 1, pp. lxxxv–lxxxvi.
67. A W Drayson, *Experiences of a Woolwich Professor during Fifteen Years at the Royal Military Academy* (London: Chapman & Hall, 1886), p. 74; [Anon.], *Some Notes on the History of the Advanced Class 1866–1926* (Woolwich: Caxton Press, 1926), p. 7.
68. [An 'Old Cadet'], 'Reminiscences of the R. M. A. in the Early "Seventies"', in F G Guggisberg, *'The Shop': The Story of the Royal Military Academy* (London: Cassell, 1900), p. 131.
69. [One of Them], 'The Story of A Batch, 1863 to 1865', in Guggisberg, 'The Shop', p. 106.
70. Guggisberg, *'The Shop'*, appendix XX, table II.

71. Paul Garcia, *The Life and Work of Major Percy Alexander MacMahon* (PhD Thesis: Open University, UK, 2006), pp. 18, 20.
72. [Anon.], *Some Notes*, pp. 2–3, 7.
73. Joseph Larmor, Letter to *The Times*, 31 December 1929, p. 8.
74. J A Venn (comp.), *Alumni Cantabrigienses: A Biographical List of all known Students, Graduates, and Holders of Office at the University of Cambridge from the Earliest Times to 1900*, II, III (Cambridge: Cambridge University Press, 1940–1954), p. 135, where he is erroneously referred to as George Alfred Greenhill.
75. Larmor, Letter to *The Times*.
76. Gian-Carlo Rota, 'Introduction', in George E Andrews (ed.), *Percy Alexander MacMahon: Collected Papers*, I (Cambridge MA: The MIT Press, 1978), pp. xiii–xiv.
77. Larmor, Letter to *The Times*.
78. Percy MacMahon, *New Mathematical Pastimes* (Cambridge: Cambridge University Press, 1921).
79. Percy MacMahon, *Combinatory Analysis* (Cambridge: Cambridge University Press, 1915–1916).
80. Whittaker, 'Forsyth', 217.
81. John Fauvel, 'James Joseph Sylvester', in John Fauvel, Raymond Flood and Robin Wilson (eds.), *Oxford Figures: 800 Years of the Mathematical Sciences* (Oxford: Oxford University Press, 2013), pp. 233, 235.

5 MATHEMATICIANS IN AN AETHEREAL WORLD

The evidence presented so far has shown that in pure mathematics Britain was an international backwater during most of the nineteenth century, and not even Arthur Cayley, generally considered to be the outstanding Victorian pure mathematician, could rival the finest of the Continentals. However, this chapter will show that in applied mathematics and theoretical physics it was a different story, and that pure mathematicians suffered from the contrast.

By the 1830s, the generalist tradition of natural philosophy and mixed mathematics, having reigned at Cambridge almost since the death of Newton in 1727, had at last begun to crystallise into more distinct disciplines; so although 'pure mathematics' had been a description of certain aspects of mathematics for more than two centuries, the pure mathematician, like the scientist, was a new arrival on the British scene. This is illustrated by Figure 5.1, in which the prevalence in printed books of the terms 'pure mathematics' and 'pure mathematician(s)' is compared over time; it is evident that the former term has a long history, whereas the latter was unknown before about 1830, and it is no coincidence that this was also when the word 'scientist' was coined. Just as it is a mistake to assume that because there was science there were also scientists, so it is to assume that because there was pure mathematics there were also pure mathematicians.

What occurred was not merely a change or updating of terminology; rather, it arose from a slow transformation of the natural philosophers of old into specialists for whom philosophy and religion could not subvert a scientific argument – that is, one based on

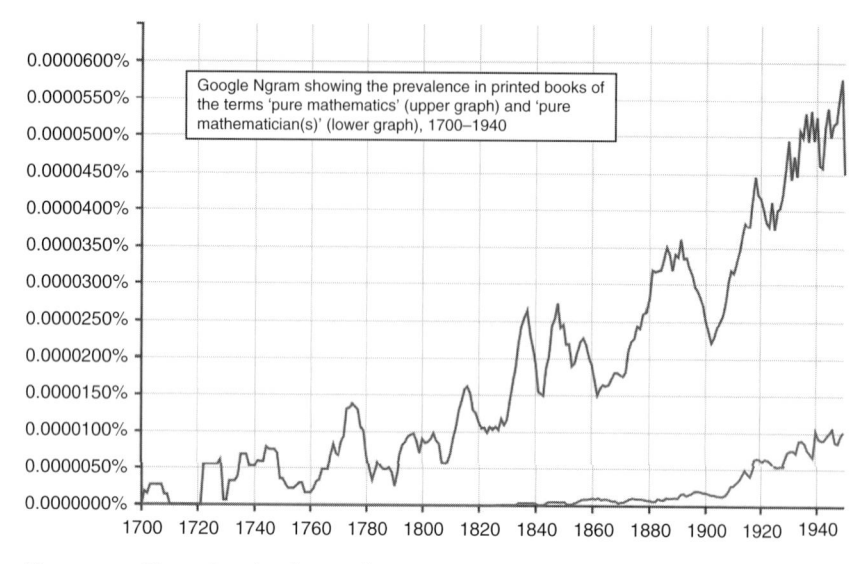

Figure 5.1 Chart showing how references to 'pure mathematician(s)' began only in the 1830s, compared to the much earlier use of 'pure mathematics'.

observation, experiment, evidence and logic – when it came to unravelling the mysteries of creation and applying that knowledge to the solution of practical problems. This trend towards specialisation continued throughout the nineteenth century, and was also evident in the universities, where the new breed of scientist was increasingly to be found in dedicated departments that reflected their specialisms. In mathematics, those mathematicians who continued to practise in the former tradition of mixed mathematics began to make advances that, on the one hand, would help to revolutionise our understanding of the physical universe, and on the other would make some of the most important contributions to Britain's international pre-eminence in industry and commerce. At the same time, pure mathematicians were attempting to promote and justify their discipline as a worthy object of mental endeavour, notwithstanding its lack of utility. They therefore had the task of distinguishing themselves in a positive manner from mathematicians who worked to forward the applied sciences, a group that included some of the world's most distinguished scientists, with a public reputation that the pure mathematicians were unable to match.

Probably the most significant theoretical and mathematical developments during this time were in light, electricity and magnetism, not only on account of utility, but also because of the ramifications that the theories had for a greater understanding of the nature of matter, space and time, which led ultimately to the special and general theories of relativity advanced by Albert Einstein during the first two decades of the twentieth century. Nevertheless, the modern investigation of these phenomena had begun almost two centuries before the Victorian era, and so we give first an outline of the historical background, before moving on to consider the achievements of British applied mathematicians and mathematical physicists.

In the first half of the seventeenth century, René Descartes proposed to replace the Aristotelian scholasticism that then prevailed with a new basis for natural philosophy, for he had concluded that everything he had been taught concerning the world was open to doubt. However, he was impressed by the certainty of mathematicians' demonstrations and, in emulation, sought some indubitable elementary principle from which he could build up by rational thought a complete system of knowledge. The principle on which he settled was 'Cogito, ergo sum', or 'I think, therefore I am' – perhaps the most famous phrase in philosophy – from which he proceeded to deduce, amongst other things, the underlying nature of the created world and the causes of natural phenomena, including mysterious 'actions at a distance' such as magnetism and the transmission of light and heat, all of which he explained in great detail in his *Principles of Natural Philosophy*.

The scholastics, following Aristotle, had postulated four primary qualities of hotness, coldness, wetness and dryness, which in various combinations gave rise to the four elements of earth, air, fire and water, out of which all matter was constituted; in addition, there were various secondary qualities such as colour, taste and odour. Descartes, on the other hand, sought a mechanistic explanation of phenomena, which he had not found in the Aristotelian doctrine. Instead, he asserted that the whole created Universe was an indefinitely large plenum, filled with a single substance that was sometimes perceptible as matter, and sometimes imperceptible; in the latter case, we mistakenly referred to empty space, or vacuum.[1] This substance consisted of tiny, infinitely divisible corpuscles, and variations in their size, shape and motion were sufficient to explain all the scholastic qualities, along with the phenomena belonging to sciences such as physics, chemistry and astronomy. Consequently, he provided an interface between medieval and modern science: on the one hand, like the scholastics he relied on logic to build up his system, but on the other hand, he abandoned metaphysical qualities, and thus paved the way for mechanistic systems such as that of Newton, which was much more successful.

According to Descartes, the corpuscles with which the Universe was filled came in three varieties, the distinction lying in their fineness and the rapidity of their motions. The first kind, the finest, were also the most rapid, and these formed the heavenly bodies; the second kind, somewhat coarser and slower, comprised the corpuscles that are imperceptible to us, and filled what was mistakenly thought of as empty space; and things here on Earth were made from the third kind, the coarsest and slowest. The circumstance that the Universe was a plenum required the motion of corpuscles to be in closed loops, because otherwise there would be an infinite chain of movement, as the place vacated by one particle was instantly filled by another. This consideration led Descartes to formulate his celebrated theory of vortices, in which corpuscles swirl around a centre, as in a whirlpool; and with the Sun as such a centre there was then a natural explanation for the otherwise inexplicable fact that all the planets orbit the Sun in the same direction: they are impelled by the corpuscles. He then explained the nature of light: at the surface of the Sun the complex interactions of the first and second kind of corpuscles resulted in the latter transmitting pressure throughout the plenum, and when this pressure impinged on our retina we perceived it as light.[2] Descartes invoked similar arguments to explain colours, the laws of reflection and refraction, magnetism, gravity and many other natural phenomena.

Because Descartes's theory of physics was very ambitious in its aims, it was sometimes inconsistent and illogical; but being entirely mechanical it had adherents for several decades, particularly amongst those who found abhorrent the idea of 'action at a distance', whereby one body could influence another body across vast tracts of empty space. Nevertheless, there were observations that the theory had difficulty in explaining, such as diffraction (which causes the edges of shadows to be diffuse, rather than precise), and Newton's rings (which are bands of colour seen when a lens is placed on a flat glass plate). The need for a theory to account for such phenomena inspired Robert Hooke, one of the founders of the Royal Society, to suggest after many experiments that our perception of light was due to pulses or waves propagated by the light source through a homogeneous medium (the *aether*); and he added that these pulses were at right angles to the direction of travel.[3] This angered Newton, who had previously provided the Royal Society with a 'New Theory about Light and Colors', in which he considered the possibility that a light source emitted particles that we perceived as light.

Hooke was very critical of Newton's conclusions, ascribing to him a belief in the hypothesis that light is a body, but Newton objected strongly, saying: "'Tis true, that from my Theory I argue the *Corporeity* of Light; but I do it without any absolute positiveness ... and make it at most but a very plausible *consequence* of the Doctrine [his theory of light and colours], and not a fundamental *Supposition*.' He continued:

> But I knew, that the *Properties*, which I declar'd of *Light*, were in some measure capable of being explicated not only by that, but by many other Mechanical *Hypotheses*. And therefore I chose to decline them all, and to speak of *Light* in *general* terms, considering it abstractly, as something or other propagated every way in streight lines from luminous bodies, without determining, what that Thing is ...

As this shows, Newton was not particularly interested in exploring hypotheses as to the nature of light, preferring to investigate its properties and behaviour; but he was clear that the corporeity of light did entail vibrations in the aether: 'For, assuming the Rays of Light to be small bodies, emitted every way from Shining substances, those, when they impinge on any Refracting or Reflecting superficies, must as necessarily excite Vibrations in the *æther*, as Stones do in water when thrown into it.'[4] This heated exchange of views between Hooke and Newton established in opposition the two hypotheses that were to dominate the arguments thereafter: the pulse, wave, undulatory or vibratory theory, in which light is a disturbance propagated through the aether, and the particle or corpuscular theory, in which light consists of 'small bodies' emitted by the light source.

In 1676 the Danish astronomer Ole Römer determined from observations of the moons of Jupiter that light must have a finite speed; and although astronomical distances were not properly understood he was nevertheless able to make an approximation of the time taken for the Sun's light to reach the Earth. However, his new data, and the advances that Christian Huygens had made in the theory of vibrations (which included substituting longitudinal waves for Hooke's transverse waves),[a] were not conclusive as to which of the two theories – wave or particle – was correct. A flow of new observations, such as that of double refraction, made the situation more rather than less uncertain; but in 1726 James Bradley, the Savilian Professor of astronomy at Oxford, discovered aberration, in which the altitude of a star above the horizon as it crosses the meridian shows a slight deviation that varies with the season of the year, and therefore with the direction of the Earth's travel in its orbit around the Sun. This effect was rightly explained by Bradley as being due to the finite speed of light, of which he was able to make a reasonably accurate estimate; and consideration of the possible means by which aberration could occur tended to support the particle theory, which remained the theory of choice for the rest of the century. However, it could not explain all the phenomena of light, so the wave theory continued to have adherents.

Following the death of Newton in 1727, a feature of the manner in which natural philosophers thought about the aether, light and 'actions at a distance' such as magnetism, is the extent to which it combined metaphysics, mathematical models and imaginary mechanical models. The question as to whether, and if so in what sense, the hypothesised entities and forces that were invoked as explanations really existed might seem to be peripheral, because the need was for accurate systems of definitions and equations that were consistent with, and could be used to predict, observed behaviour. Nevertheless, it was not enough that such a mathematical model gave the right answers; to have confidence in it, the mathematical processes also had to be understood as being analogous to natural processes that could give rise to the phenomena. There was also a psychological need, particularly amongst the Victorians, to imagine hypothesised entities and forces in mechanical terms, so that the mathematical system was underpinned by a mechanical model to which the principles of classical mechanics could be applied.[b] The attraction of Descartes's system had been that it offered a purely mechanical explanation of so many diverse phenomena.

During the eighteenth century, natural philosophers attempted to model and explain the behaviour of light, magnetism and electricity,

---

[a] Longitudinal waves are those in which the pulses are in the direction of travel; transverse waves are those in which the pulses are at right angles to the direction of travel.
[b] Classical mechanics is the branch of mechanics that derives from Newton's laws of motion.

each of which presented its own difficulties. Waves can be propagated through a medium only because of its elasticity,[c] of which different phenomena were thought to require different degrees; and this implied separate aethers. However, at the close of the eighteenth century Thomas Young, a London doctor and experimenter, challenged this view with arguments to suggest that the properties of the electrical and luminiferous aethers would be very similar, although he stopped short of saying that they were one and the same.[5] Furthermore, he successfully used the manner in which waves can interfere with one another to explain some of the phenomena of light that had confounded the corpuscular theory. This swung the balance in favour of the wave theory, but the matter was by no means settled, and the corpuscular theory gained a boost from the discovery of polarisation, which occurs when light is reflected or passes through certain crystals, and which Young was unable to explain.

In spite of this setback, the triumph of the wave theory was at hand, because problems for the corpuscular theory had arisen from new investigations into stellar aberration and its effect on the refraction of starlight in glass, which had to be taken into account in the design of telescopes.[6] Young realised that almost all the difficulties encountered by the wave theory could be overcome if it was modified so that the vibrations were no longer considered to be longitudinal but instead were transverse, with undulations at right angles to the direction of travel[7] – which is what Robert Hooke had said a century-and-a-half earlier. The French physicist Augustin-Jean Fresnel then elaborated this idea with such success that the corpuscular theory rapidly fell into disfavour, but there were still problems concerning the elasticity that the luminiferous aether had to possess in order to propagate transverse waves. Given the frequency of the vibrations that constituted light, a consequence of the wave theory was that the aether had to be considered not simply as an elastic medium such as water, but as an elastic solid, and to this there were serious objections that were not dispelled by the theory's success in accounting for phenomena that had hitherto resisted any explanation. The first and most obvious objection was that we have no perception of living in and moving through any kind

---

[c] In physics, elasticity refers to the tendency of the constituent particles of a substance to return to their original positions after a small displacement, due to the strength of the bonds between them. The weakest bonds are in gases; those in liquids are stronger; the strongest are in solids. Thus solid substances, with the strongest bonds, are also the most elastic.

of solid, and the Earth seems quite unimpeded by its presence; but this was not as decisive as one might suppose, because some substances, such as pitch, do have the propagating power of an elastic solid while allowing solid objects to move through them, so it could not be argued that such a thing was unknown in nature.[8]

That brings us to the second objection to the elastic solid that Young's theory required, and although not obvious it was considerably more damaging for the devisers of mechanical models. James MacCullagh was an Irish mathematician who in the 1830s made new and comprehensive researches into the mathematical representation of the behaviour of light, and concluded that the aether necessarily propagated transverse waves in a manner such that no mechanical analogy could be conceived. He then went further, and in 1839 described a mathematical model that successfully accounted for all the mathematically expressed behaviour of light by postulating an aether with 'rotational elasticity';[9] its drawback was the impossibility of imagining such an aether in mechanical terms. So although MacCullagh's aether fulfilled the mathematical requirements, it was largely ignored by the Victorians – wedded as they were to mechanical analogies – until it was resurrected nearly half-a-century later by Lord Kelvin; MacCullagh himself committed suicide in 1847. Contrary to what might have been expected, his findings did not dent confidence in the aether, because the practical success of Young's transverse waves, coupled with the impossibility of supposing that a wave motion could exist without there being anything to carry the wave, reinforced the conviction that the aether did exist in some form or another, a conviction that became almost the fortieth Article of Religion. Despite this, the lack of a mechanical model was very unsettling for British mathematicians, particularly those trained in the Cambridge tradition – which meant most of them.[10]

Nowadays we believe that light is an electromagnetic phenomenon, but in the opening decades of the nineteenth century investigations into the connection between the various manifestations of electricity and magnetism were only just beginning, with no thought that light might be one of them. Notwithstanding the perceived importance of the mathematical representation of phenomena, at that time the most influential experimentalist working on electricity and magnetism was Michael Faraday, who never went to university, and could not be described as a mathematician. Nevertheless, it is difficult to understand the contributions made to applied science by Victorian mathematicians without focussing first on his achievements.

Faraday was born in 1791 at Newington Butts, near London, was largely self-educated, and after leaving school served for several years as an apprentice to a bookbinder. While working there he read some of the books that were brought for binding, developed a particular interest in science, and attended the public lectures given at the Royal Institution of Great Britain by Sir Humphrey Davy, the famous chemist, who was so impressed by Faraday's diligence that in 1813 he engaged him as an assistant at the Institution's laboratory. Faraday spent the rest of his working life there, and it was there that he conducted the investigations, first in chemistry and then in magnetism and the various forms of electricity, that brought him worldwide fame. However, his lack of formal education meant that he had no great facility in mathematics; rather, his reputation rested on his experimental work and the physical theories that he developed, for it was these that provided inspiration for subsequent generations.

When Faraday began his researches, the Danish physicist Hans Christian Ørsted had recently discovered that an iron bar would act as a magnet if it had wound around it a wire through which an electric current was flowing. Faraday then succeeded in harnessing this phenomenon to produce rotatory motion – the basis of the electric motor – and published his results in 1822, when he was still a 'chemical assistant' in Davy's laboratory.[11] Unfortunately for Faraday, although his work was a clear advance, he failed to give due acknowledgement to others working with the same aims, and particularly offended William Wollaston, a very eminent British chemist. To remedy the situation, Faraday had to issue a somewhat grovelling apology; as with most advances in science, his achievements did not emerge from nowhere.[12]

For the next few years Faraday was unable to devote much time to these researches, but in 1831 he made what is perhaps his greatest discovery, namely that of electromagnetic induction; in the report of his experiments that he gave to the Royal Society he summarised his results with the laconic comment: 'Here therefore was demonstrated the production of a permanent current of electricity by ordinary magnets.'[13] He had demonstrated the principle of the dynamo, which allows a current to be generated in a circuit if the conducting wire and a magnet are constantly in relative motion; half-a-century later, this led to the introduction of the large steam-powered generators that were required for the widespread distribution of electrical power.

Faraday was no mathematician, but he was more than simply an experimenter; he thought long and deeply about the principles that lay behind electrical and magnetic phenomena, and his theoretical work was to prove very influential in the mathematics that was later to be done by others. In particular, he introduced the idea of magnetic *lines of force* in space; these were non-intersecting closed curves passing through the poles of a magnet, such that at each point on a line the direction of the magnetic force was along the tangent to the curve. In this way, every point subject to the influence of a magnet had associated with it not only the strength of the force at that point, but also the direction in which it acted; the two together constituted what later came to be described as a vector.[14] Later, in the 1840s, he tried to expand the remit of lines of force by suggesting that they vibrated, filled the whole of space, and could explain all radiation phenomena, including light.[15] Although lines of force did not displace the aether in the minds of his contemporaries, they brought with them the concepts of electric and magnetic fields, which were to figure prominently in the work of future mathematical physicists. Faraday's multifarious experiments, inventions and theories had given him worldwide fame, as can be gauged by the extent to which his name is still invoked today: in the international unit of capacitance (the *farad*), the Faraday constant, Faraday's laws of electrolysis and induction, Faraday rotation and the Faraday cage. Against conventional opinion, Faraday believed that all forms of electricity were fundamentally the same, and this belief may have been influenced by his deep religious convictions. He belonged to the Sandemanians, a Presbyterian sect that adopted some of the Bible's teachings more strictly than did mainstream Presbyterianism, and his faith gave him a predisposition to believe in the underlying unity of creation.

Among the Victorians, adopting the persona of the scientist never precluded strong religious belief, and this trait was shared by James Clerk Maxwell who, more than anyone else, was responsible for clothing the experimental results of Faraday and others in the garb of mathematics. A Scotsman, Maxwell was born into a middle-class family in 1831, and unlike Faraday had a scientifically conventional education, attending the University of Edinburgh for three years before beginning studies at Cambridge for the Mathematical Tripos. In 1854 he emerged as second wrangler and a joint first Smith's prizeman, and was for two years a Fellow of Trinity College; then, after 15 years at the Universities of Aberdeen and London, he returned to Cambridge in 1871 as the first Cavendish Professor of experimental physics and director of the Cavendish Laboratory, positions that he retained until his unexpected death in 1879.

During his years away from Cambridge he had developed a particular interest in the kinetic theory of gases, which presents gas as being comprised of molecules that move with varying speeds and collide with one another in a random fashion; as Maxwell hastened to point out, these molecules are very different from the molecules (or corpuscles) of Descartes.[16] In the kinetic theory, the properties that we measure experimentally as being the temperature and pressure of a volume of gas are related to the number and speed of the molecules, so in principle the properties of the gas could be determined dynamically – that is, by considering the individual behaviour of each molecule, just as one might consider the behaviour of billiard balls ricocheting on a billiard table. Of course, this is not possible in practice, so statistical methods that invoke the average behaviour of the molecules are used to support experimentally determined gas laws that treat a given sample of gas as a single entity; for example, if there is a fixed mass of gas in a container of constant volume, then the temperature of the gas (measured on the absolute scale) is proportional to the pressure that it exerts on the container.

When Maxwell began his investigations, the kinetic theory had already been significantly developed by the German physicist Rudolf Clausius, who in the process had formulated the second law of thermodynamics, which Maxwell explained:

> One of the best established facts in thermodynamics is that it is impossible in a system enclosed in an envelope which permits neither change of volume nor passage of heat, and in which both the temperature and pressure are everywhere the same, to produce any inequality of temperature or of pressure without the expenditure of work.

In 1859 Maxwell began to publish a series of papers in which he developed a mathematical treatment of the dynamical and kinetic theories; these papers established his reputation as one of the foremost mathematical physicists of the day. His popular textbook *Theory of*

Heat was published in 1871, and when discussing thermodynamics he introduced what soon became a famous argument suggesting that the second law was open to challenge. The law only holds true (he said) because we are unable to follow the individual motions of the molecules that comprise the gas; but if our powers were enhanced so that we could do this, then the law could be contravened. Referring to the envelope (or vessel) that he had previously described, he continued:

> Now let us suppose that such a vessel is divided into two portions, A and B, by a division in which there is a small hole, and that a being, who can see the individual molecules, opens and closes this hole, so as to allow only the swifter molecules to pass from A to B, and only the slower ones to pass from B to A. He will thus, without expenditure of work, raise the temperature of B and lower that of A, in contradiction to the second law of thermodynamics.

'Work', here, refers to work done on the gas, not to any work that may be done in measuring the speed of the molecules, or opening and closing the hole. The imaginary being – who subsequently became known as 'Maxwell's Demon' – demonstrated, in Maxwell's view, that what is believed to be a fundamental law of physics is not necessarily such, if our belief in it stems from our inability to perceive and manipulate the ultimate constituents of matter. Maxwell then mused as to whether, and if so to what extent, scientific concepts that had been derived using dynamical methods can be applied to 'our actual knowledge of concrete things, which, as we have seen, is of an essentially statistical nature...'. The implications of this argument were wide-ranging, and 'Maxwell's Demon' is probably the only imaginary being who has made a significant contribution to scientific thought.[17]

Maxwell's work on heat and the kinetic theory of gases had secured his reputation as a mathematical physicist, but his worldwide fame derived, and still derives, from his theoretical work in electromagnetism. From his days as an undergraduate at Cambridge he had been a keen student of Faraday's theories, and had drafted a paper 'On Faraday's Lines of Force' before taking the Tripos, although it was not presented in public until 1855.[18] During the next few years he absorbed all the work on electromagnetism that had been done by others, and began to cast into a mathematical form the experimental results that had been obtained during the previous half-century.

Far-reaching though his conclusions were to be, he was a product of his age: like most Victorians he accepted the necessity of an aether and was an enthusiast for mechanical models, which led his lifelong friend Peter Guthrie Tait to say of him that he 'preferred always to have before him a geometrical or physical representation of the problem in which he was engaged, and to take all his steps with the aid of this: afterwards, when necessary, translating them into symbols'.[19] This method certainly served Maxwell well, for by 1873 he had written and published his comprehensive *A Treatise on Electricity and Magnetism*, which gave a full account of the theory that he had developed during the preceding 20 years, and has since become one of the classics of science.

Nowadays Maxwell's name is most familiar from the differential equations – universally known as 'Maxwell's equations' – that give a complete mathematical description of the interplay between electric and magnetic phenomena, and remain a staple of university physics courses; these equations, and the conclusions that he derived from them, were to determine the whole course of electromagnetic theory and practice. Central to Maxwell's theory was the concept of the *field*, which led Albert Einstein to remark: 'The most fascinating subject at the time that I was a student was Maxwell's theory. What made this theory appear revolutionary was the transition from action at a distance to field as the fundamental variables.'[20] Maxwell defined the field as being 'that part of space which contains or surrounds bodies in electric or magnetic conditions. It may be filled with any kind of matter, or we may endeavour to render it empty of all gross matter ... '.[21] Formerly, one such body was deemed to act directly, albeit mysteriously, on another distant body; but, in the new dispensation, at every point of a field there was deemed to be electric or magnetic influence, the magnitude and direction of which could be described mathematically. It was this focussing of attention on the mathematical description of space that so enthused Einstein and his fellow students. In particular, an electric field could be induced in certain materials – *dielectrics* – that acted as insulators, and in such a case the molecules of which the dielectric was composed were subject to an electromotive force.[d] Maxwell said:

---

[d] An example is a simple capacitor, in which two parallel metal plates, one charged positively and one negatively, are separated by an insulating dielectric. Although no current can flow between the plates, the charges on them induce an electric field in the dielectric.

In a dielectric under induction, we may conceive that the electricity in each molecule is so displaced that one side is rendered positively, and the other negatively electrical, but that the electricity remains entirely connected with the molecule, and does not pass from one molecule to another. The effect of this action on the whole dielectric mass is to produce a general displacement of the electricity in a certain direction. This displacement does not amount to a current, because when it has attained a certain value it remains constant, but it is the commencement of a current, and its variations constitute currents in the positive or negative direction, according as the displacement is increasing or diminishing.[22]

This was his famous *displacement current*.

Maxwell's original equations, based as they were on experiments undertaken by others, specified the induction of an electric field by a magnetic field, and the induction of a magnetic field by an electric current; but there was no evidence to suggest that a magnetic field was induced by an electric field, and so there was at first no term in the equations representing such an effect. Nevertheless, his conclusion that an electric field in a dielectric must result in a displacement current suggested to him that a varying electric field should induce a magnetic field, as did an electric current in a conductor, and so he added an appropriate term to his equations.[23] Although these showed that any induced magnetic field would be very weak, which could explain why its existence had not been detected in experiments, there was an immediate theoretical consequence, for the equations were now symmetrical: a varying electric field induced a varying magnetic field, and vice versa. Therefore, after the initial induction, there would be a rapid succession of electric and magnetic fields as each induced the other; and then it was but a short step (at least for Maxwell) to derive wave equations suggesting that these successive electric and magnetic fields were propagated as electromagnetic disturbances, or waves, in the aether.

The paper in which he introduced the displacement current was published in 1862, and it contained a further result of great interest, for a consequence of his theory was that the speed, in millimetres per second, with which electromagnetic waves were propagated should equal 'the number by which the electrodynamic measure of any quantity of electricity must be multiplied to obtain its electrostatic measure'; and this the German physicists Wilhelm Weber and Friedrich Kohlrausch had experimentally determined to be 310,740,000,000. Maxwell then compared this figure with the best experimental determination of the speed of light that had so far been made, which was by Hippolyte Fizeau and yielded 314,858,000,000 millimetres per second. The theoretical speed of electromagnetic propagation agreed so well with this result that, Maxwell said, 'we can scarcely avoid the inference that *light consists of the transverse undulations of the same medium which is the cause of electric and magnetic phenomena*'.[24] This contradicted the widespread belief that there were at least two aethers; and three years later he used the same figures and arguments to offer one of the most significant and far-reaching conclusions in nineteenth-century physics: 'The agreement of the results seems to show ... that light is an electromagnetic disturbance propagated through the field according to electromagnetic laws.'[25] Not only was there a single aether: there was also only one phenomenon, that of electromagnetism. Maxwell used the cautious phrase 'seems to show' because no equation could *prove* that light is an electromagnetic disturbance, and in any event electromagnetic disturbances in the aether were at that time theoretical only, awaiting experimental confirmation. However, the 'agreement of the results' demonstrated beyond reasonable doubt that if Maxwell was right and electromagnetic disturbances *were* propagated through the aether in the manner described by his equations, then light was an electromagnetic disturbance.

Maxwell's theory illustrates well how suggestive a good mathematical model can be, because when he died in 1879 at the age of 48 electromagnetic waves were still only an unverified consequence of his theory. During the next decade experimentalists tried to demonstrate their existence in the laboratory, but it was not until 1889 that Heinrich Hertz succeeded in detecting them, and showing that their properties were as predicted by Maxwell.[26] Two developments of almost limitless significance flowed from this. One was radio communication; the other was a fundamental rethinking of the relationship between time, space and motion, for Einstein acknowledged that 'The special theory of relativity owes its origin to Maxwell's equations of the electromagnetic field.'[27] Consequently Maxwell enjoys a well-deserved place in the very first rank of mathematical physicists.

Maxwell was also indirectly responsible for one of the most famous experiments in the history of physics; which brings us once again to the vexed question of the nature of the aether. The corpuscular theory of light had received its *coup de grâce* in 1850, when an experiment by Fizeau and Foucault demonstrated conclusively that the speed of light was less in water than in air, a result that was predicted by the wave theory but contradicted by the corpuscular theory. So there was now no turning back from the belief that light was propagated by transverse waves, but mathematicians and physicists were still haunted by the question: waves in what? Common sense demanded a medium, and the model-making mind-set of the Victorians demanded a medium for which a plausible mechanical analogue could be devised; but the work of MacCullagh had shown that this demand was probably impossible to fulfil. Although for some mathematical physicists the existence of the aether remained a given well into the twentieth century, much that was paradoxical followed from it, as the mathematician William Kingdon Clifford had not hesitated to make clear in a popular lecture published in 1879, the year of his death. In 1867 he had been second wrangler and second Smith's prizeman at Trinity College, and was one of the most innovative thinkers of the day on such matters. Referring to the 'shake' (or vibration) in the aether that was perceived as the light of a gas lamp, he said:

> In order to carry a shake such as this it is necessary to suppose that the luminiferous ether is not a fluid like water, but that it is a solid, something like a piece of jelly. It is an exceedingly difficult thing to conceive how there should be a separate substance filling all space, and filling up all the interstices between different molecules of bodies, and which yet leaves us able to walk about in the midst of it as we do. But that is the truth. There is a solid substance not made up of the same molecules as ordinary matter, but which is such that all these molecules move about in it, and when they shake they produce waves of disturbance which spread round in this solid substance in all directions, and these waves are what we call light.[28]

It is not recorded what his audience made of this. As we shall see later, Clifford was not as convinced about the existence of the jelly-like luminiferous aether as his lecture might suggest; nevertheless, it was difficult to imagine how light could be propagated without it.

Clifford's reference to jelly highlighted a major difficulty – to which we have already alluded – concerning the elasticity of the aether:

the speed and frequency of the transverse waves that constituted light required a highly elastic medium, which implied a solid; but if (as was generally believed) the aether offered no resistance to the Earth in its orbit around the Sun, then it had to be a fluid – two seemingly incompatible demands. The comparison with jelly (which fortunately cannot be taken very far) had been suggested by Stokes,[29] because, as Larmor explained, 'The coexistence of fluidity on a large scale with perfect elasticity on a small scale [is illustrated] by the ordinary phenomena of pitch or glue, passing on to a limit through jellies of gradually diminishing consistency, until perfect fluidity is reached...'[30]

The riddle of the solidity or fluidity of the aether was intimately bound up with another difficulty, namely the relation between the aether and the heavenly bodies. If the aether offered no resistance to the orbiting Earth – if, in Thomas Young's bucolic simile, the aether 'pervades the substance of all material bodies with little or no resistance, as freely perhaps as the wind passes through a grove of trees'[31] – then the Earth and the aether would be in relative motion; but if there was 'aether drag', with the orbiting Earth dragging the nearby aether with it, then at the Earth's surface there would be no relative motion. The answer mattered for the accuracy of terrestrial and astronomical observations, because if the Earth was moving through an aether at rest (so that the aether was moving relative to the Earth), then one would expect the speed of light at the Earth's surface to vary according to whether or not it was propagated in the same direction as that in which the orbiting Earth was travelling (the Earth's rotation being an additional complication); whereas if the Earth dragged the terrestrial aether with it, then at the Earth's surface the two would not be in relative motion, and the speed of light would be the same in all directions. It was also possible, of course, that there was partial aether drag, but it was generally held, on theoretical grounds, that there should be none at all.[32]

In 1879 Maxwell wrote to D P Todd, of the U S Nautical Almanac Office, and remarked that any terrestrial experiment to determine the matter would involve 'a quantity depending on the square of the ratio of the earth's velocity to that of light, and that this is quite too small to be observed'.[33] This remark was seen by Albert Michelson, a young assistant at the Office, who began to consider the possibility of detecting such a quantity; and by 1881 he had devised an experiment using interferometry, a technique that analyses the interference patterns produced by the superimposition of light waves. If his apparatus yielded a positive result, as he and everyone else anticipated, then it would prove that the speed of light was different in different directions, and therefore that aether drag was non-existent, or nearly so; but what he obtained was a totally unexpected null result, which suggested that the aether drag was either complete, or so nearly complete that it yielded a difference too small to be detected.[34] Unfortunately, it was then discovered that the experiment was less precise than had been thought, but a few years later, and with Edward Morley as a collaborator, he used much more accurate apparatus to repeat several times what is now known as 'the Michelson–Morley experiment' – always with the same negative outcome. This led them to conclude that it is:

> reasonably certain that if there be any relative motion between the earth and the luminiferous ether, it must be small... But it is not impossible that at even moderate distances above the level of the sea, at the top of an isolated mountain peak, for instance, the relative motion might be perceptible in an apparatus like that used in these experiments.[35]

These experiments, far from clarifying matters, just added to the considerable confusion that then existed concerning the aether. Complete aether drag at the Earth's surface, which is what the very accurate measurements implied, was implausible, particularly as an aether that was unaffected by the passage of heavenly bodies appeared to offer the measure of universal absolute rest that was postulated by Newtonian mechanics; on the other hand, most mathematicians and physicists were not willing to contemplate the alternative of denying the existence of the aether altogether. Clearly, some means had to be found of making its existence compatible with the null result of the experiments, a feat that was performed independently by two mathematicians, the Irishman George FitzGerald and the Dutchman Hendrik Lorentz, who suggested that the difficulty could be overcome if a moving body became shorter in the direction of motion by an amount determined by the equation $L = L_0 \sqrt{1 - \frac{v^2}{c^2}}$, where L is the contracted length, L_0 the uncontracted, v the speed of the body, and c the speed of light.[36] This, if true, would indeed explain the outcome of the experiment, but it was also a very unsatisfactory explanation because the equation is totally ad hoc, having been especially invented to save the hypothesis of the aether.

However, the alternatives were equally unpalatable, so physicists were left in a quandary that was not resolved until the next century, when Einstein developed the special theory of relativity.

Maxwell did not live to see the outcome of the Michelson–Morley experiment, but had he done so he would have had the satisfaction of knowing that his comment to Todd, although wrong, had led to one of the most significant experiments in Victorian physics. The lasting value of his work is not something that was recognised only by posterity; his contemporaries all acknowledged that he had had a profound influence on electromagnetic theory. That influence continues, although today a student might not recognise Maxwell's equations as they were first published because the modern presentation uses vector notation, which is much more compact. This recasting of the originals was largely the work of Oliver Heaviside, a British mathematician and telegraph engineer of unconventional background, who made significant advances in many aspects of practical electromagnetism as it affected telegraphy, as a result of which he was nominated seven times for the Nobel Prize in Physics – although always by the same person.[37] His name lives on because he predicted the existence of the Heaviside layer, which consists of ionised particles that reflect radio waves about 100 km above the Earth, and thereby makes possible long-distance radio communications; his prediction was not experimentally confirmed until 1923, two years before his death. He was self-educated, and said that 'I regret exceedingly not to have had a Cambridge education myself, instead of wasting several years of my life in mere drudgery, or little more.'[38] Nevertheless he was inspired to study mathematics by reading Maxwell's *Treatise*, believed that Maxwell was 'a heaven-born genius',[39] and along with the American Josiah Willard Gibbs was largely responsible for the modern theory of vectors, at a time when they were little understood and regarded with suspicion – he thought of himself as 'a prophet howling in the wilderness'.[40] Unlike Maxwell he received little public recognition in his lifetime, and without a university education he was not part of the circle of Cambridge alumni to which most significant Victorian mathematicians belonged.

In the late Victorian period, and in the face of accumulating problems, the enthusiasm of British mathematicians and physicists for inventing aethers of various kinds did not diminish; in Arthur Eddington's words, 'The nineteenth century is littered with the debris of abortive aethers – elastic solids, jellies, froths, vortex networks ...'.[41] Lord Kelvin (William Thomson) felt an especial need to imagine a mechanical model of a theory, saying in a lecture: 'I never satisfy myself unless I can make a mechanical model of a thing. If I can make a mechanical model, I can understand it. As long as I cannot make a mechanical model all the way through, I cannot understand; and that is why I cannot get the electromagnetic theory.'[42] He showed great ingenuity in devising models, and succeeded thereby in resurrecting the 'rotational elasticity' of MacCullagh's aether. Edmund Whittaker described the model, which is a good illustration of the lengths to which Victorian mathematicians were prepared to go in their quest for a mechanism:

> Suppose, for example, that a structure is formed of spheres, each sphere being in the centre of the tetrahedron formed by its four nearest neighbours. Let each sphere be joined to these four neighbours by rigid bars, which have spherical caps at their ends so as to slide freely on the spheres. Such a structure would, for small deformations, behave like an incompressible perfect fluid. Now attach to each bar a pair of gyroscopically-mounted flywheels, rotating with equal and opposite angular velocities, and having their axes in the line of the bar: a bar thus equipped will require a couple[e] to hold it at rest in any position inclined to its original position, and the structure as a whole will possess that kind of quasielasticity which was first imagined by MacCullagh.[43]

In the minds of those who, like Kelvin, demanded a mechanical model, this allowed the vortex to be established as an allowable analogy for an aether, but not just in the form of vortex networks; there were also vortex rings and vortex sponges, this last being advocated by Kelvin after he had developed his mechanical analogy and had shown in 1887 that an equation of propagation that the aether had to satisfy was also an equation of propagation in a vortex sponge. Vortex sponges enjoyed a long life, and were still being commended by Whittaker in 1910.[44]

The foregoing account has illustrated the degree of attachment that most Victorian mathematicians and mathematical physicists felt towards the concept of the aether, preferably one for which a plausible mechanical analogy could be devised, yet at least one mathematician

---

[e] A couple consists of two parallel forces, equal in magnitude but opposite in direction, that act at different points of a body and so impart a turning motion.

had doubts as to whether it was necessary or helpful to make such an assumption. An alternative line of thought, expounded by William Kingdon Clifford, involved abandoning the purported Euclidean structure of the Universe and replacing it with 'curved space'. Non-Euclidean geometry – the fundamental mathematical advance to which 'curved space' refers – was destined to play a role in the general theory of relativity formulated by Einstein during the first two decades of the twentieth century, a theory that provided the most comprehensive revision of our modelling of the Universe since the *Principia Mathematica* of Newton. The development of non-Euclidean geometry had not been due to British mathematicians, but Clifford, writing decades before Einstein, used it to formulate hypotheses that foreshadowed some of the ideas that were to be found in Einstein's theories.

Non-Euclidean geometry arose in Europe as a speculative exercise during the first 60 years of the nineteenth century, with Carl Friedrich Gauss, János Bólyai, Nikolai Lobachevski and Bernhard Riemann being the main participators, but it had attracted little notice in Britain until at the British Association meeting in 1869 Sylvester said in his address to Section A that:

> Riemann has written a thesis to show that the basis of our conception of space is purely empirical, and our knowledge of its laws the result of observation, that other kinds of space might be conceived to exist subject to laws different from those which govern the actual space in which we are immersed, and that there is no evidence of these laws extending to the ultimate infinitesimal elements of which space is composed.[45]

Shortly afterwards Hermann von Helmholtz discussed these ideas in *The Academy*,[46] and in 1873 Clifford published a translation of some of Riemann's papers in *Nature*,[47] following it up the next year with a series of lectures at the Royal Institution; consequently, by the mid 1870s the possibility of an alternative to Euclid had been canvassed in Britain through non-specialist media.[48] In 1879 the Cambridge astronomer Sir Robert Ball followed the trend by publishing a paper on non-Euclidean geometry in the journal *Hermathena* which, as its name suggests, was (and still is) principally concerned with philosophy and the classics.[49] The subject was given further widespread exposure by the publication of two books that are still in print. One, entitled *Flatland: A Romance of Many Dimensions*,[50] first appeared in 1884. The author was Edwin Abbott Abbott,[f] the headmaster of City of London School; he used the pseudonym 'A Square', and drew an ingenious analogy between the experiences of a creature inhabiting a two-dimensional section of our three-dimensional world, and what our own experiences might be if we were living in a three-dimensional section of a four-dimensional world. More substantial was *The Common Sense of the Exact Sciences*, a posthumous book by Clifford published in 1885, six years after his death; it sold so well that by 1907 it had reached its fifth edition.[51] All these popular and readily available expositions ensured that, in the 30 years following Sylvester's address to the British Association, the puzzling consequences of non-Euclidean geometry and 'curved space' had been presented to scientifically inclined laymen, and were no longer confined to mathematicians.

Although non-Euclidean geometry could be considered in a purely mathematical context, in practice it was inextricably linked to questions concerning the nature of the space that we inhabit, and the behaviour of bodies in it. In addressing these questions, British mathematicians had shown their strong attachment to empirical foundations by continuing to venerate Euclid's *Elements*, because of the extent to which it reflected intuition and experience. A Euclidean space at rest was also a necessary backdrop for Newtonian mechanics, which for more than two centuries was universally accepted as an explanation of how and why material bodies moved, both here on Earth and in the heavens. Therefore geometry – which did not need to be described as Euclidean, because there was none other – appeared to be intrinsic to the design of the Universe; so until the nineteenth century 'non-Euclidean geometry', literally 'non-Euclidean Earth-measuring', would have been considered a contradiction in terms.

The *Elements* opens with a series of definitions, postulates and axioms, all of which appeal to intuition and common experience to render them acceptable; they are not proved or demonstrated – rather, they provide the foundations on which the whole edifice of Euclidean geometry is erected. However, the fifth postulate (sometimes referred to as axiom 12, or the parallel postulate) was considered by many to be less satisfactory than the others in its mode of expression, although no one doubted its truth:

---

[f] His parents were first cousins, and his second given name was his mother's maiden name.

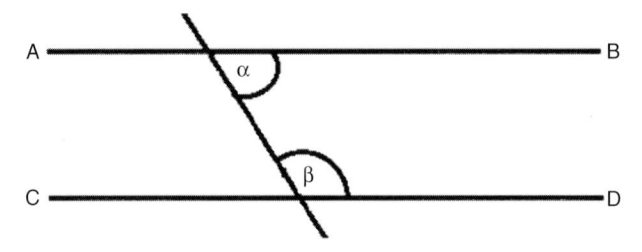

Figure 5.2 Illustration of the controversial 'parallel postulate' from the *Elements of Euclid*: if the angles α and β sum to less than two right angles, then the lines AB and CD will meet on the right.

If a straight line meet two straight lines, so as to make the two interior angles on the same side of it taken together, less than two right angles, these straight lines being continually produced, shall at length meet on that side on which are the angles which are less than two right angles.[52]

This translation is taken from Isaac Todhunter's 1862 edition of Euclid, which was popular with schoolteachers because every line of a proof is referenced back to an axiom, postulate, definition or earlier theorem, and so Victorian schoolchildren and students would have been very familiar with it. The parallel postulate was infamous because it is long, convoluted and (in the opinion of many) unintuitive, so in this respect it differs from the other postulates and axioms, which are brief and immediately convey their meaning; for example, 'All right angles are equal to one another.' (See Figure 5.2.) Furthermore, no use is made of the parallel postulate until the proof of proposition 29 in Book I, a comparatively advanced stage, and this proposition is then applied almost immediately in proposition 32 when proving an essential property of triangles, namely that the interior angles of any triangle are together equal to two right angles. This gives the fifth postulate an ad-hoc nature that the others do not have, and suggests that it was introduced as the simplest way of making possible the proof of proposition 32.

Many mathematicians thought that the fifth postulate sullied the perfection of Euclid's formalisation, and engaged in numerous but fruitless attempts to show that it was a consequence of the other axioms, postulates and definitions – in other words, a theorem – and was therefore unnecessary. The great French mathematician Adrien-Marie Legendre thought that he had succeeded, and included a 'proof' in successive editions of his popular book *Elements of Geometry*, but the 'proof' contained a subtle flaw.[53] Furthermore, for an extreme case in which the sum of the two interior angles is less than two right angles by a very small quantity, it is possible that the most careful investigation would pronounce the two lines to be a constant distance apart; yet the fifth postulate requires us to accept that those lines will meet, although necessarily at a great distance. In short, the postulate seems less obvious than the others, which are intuitively reasonable and pronounce on local experience, but its truth was unquestioned, so to deny it was a totally speculative exercise.

In the eighteenth century, mathematicians devised axioms that are logically equivalent to the fifth postulate, but simpler and more intuitive; perhaps the best known, sometimes referred to as Playfair's axiom, says that if a straight line and a point not on the line are given, then there is precisely one line that passes through the point and is parallel to the given line. The self-evident nature of this just reinforced the conviction that whatever followed from denying the fifth postulate (or Playfair's axiom) could not relate to the space of our experience, and so was not geometry, by definition. Conversely, if as a speculation the truth of the fifth postulate *was* denied, then two types of replacement geometry were possible, and so observation and experiment would have been required to determine which of them was 'true'.

A further consideration was that Immanuel Kant – often seen as the most influential philosopher of modern times – had been concerned to show the possibility of knowledge that was both synthetic (that is, not derived simply from an analysis of the concepts), and a priori (that is, not the product of experience). Although these may seem contradictory requirements, Kant believed that they were manifested in Euclidean geometry; so if it turned out that the geometry of the Universe could be determined only by empirical methods, and was therefore not a priori, it weakened his epistemological arguments and his theory of mind. After it became widely accepted that a non-Euclidean geometry *could* be the geometry of our experience, Bertrand Russell, in his fellowship dissertation, attempted a partial restoration of Kant's epistemology by arguing that projective geometry was a priori, and could fulfil the role that Kant had given to

Euclidean geometry.[g] The dissertation itself is now lost, but Russell worked it up into his first book.[54]

During the second half of the nineteenth century, commitment to the absolute truth of the fifth postulate began to waver. The former natural philosophers, newly reinvented as 'scientists', found it acceptable to consider the possibility that the postulate might be, as Riemann had suggested, only an approximation so close to the truth as to render any deviation undetectable, with the consequence that the intrinsic geometry of the Universe might not be precisely Euclidean. This hypothesis led to speculation concerning 'curved space' and the behaviour in it of light and material bodies, the unforced motion of which was, according to the Newtonian theory, in Euclidean straight lines. However, if Euclidean behaviour was only approximate, then the motion of light and material bodies would have to be described in accordance with the precepts of the true geometry, whatever that happened to be.

Sylvester, in a footnote added to the printed version of his address of 1869 to the British Association, said that:

> It is well known to those who have gone into these views that the laws of motion accepted as a fact suffice to prove in a general way that the space we live in is a flat or level [Euclidean] space ... our existence therein being assimilable to the life of the bookworm in an *unrumpled page*: but what if the page should be undergoing a process of gradual bending into a curved form? Mr W. K. Clifford has indulged in some remarkable speculations as to the possibility of our being able to infer, from certain unexplained phenomena of light and magnetism, the fact of our level space of three dimensions being in the act of undergoing in space of four dimensions ... a distortion analogous to the rumpling of the page to which that creature's powers of direct perception have been postulated to be limited.[55]

---

[g] Projective geometry deals only with geometrical properties that are invariant under a projection, such as the order of points on a line, or whether two lines intersect, or whether a line is straight; but it does not differentiate between, say, a circle and an ellipse. Therefore it is often thought of as being more fundamental than Euclidean and non-Euclidean geometries.

The next year Clifford put these speculations forward in a paper entitled 'On the Space-Theory of Matter' presented to the Cambridge Philosophical Society, in which he said:

(1) That small portions of space *are* in fact of a nature analogous to little hills on a surface which is on the average flat; namely, that the ordinary laws of geometry are not valid in them.
(2) That this property of being curved or distorted is continually being passed on from one portion of space to another after the manner of a wave.
(3) That this variation of the curvature of space is what really happens in that phenomenon which we call the *motion of matter*, whether ponderable or ethereal.
(4) That in the physical world nothing else takes place but this variation, subject (possibly) to the laws of continuity.[56]

Such were the views, already publicly expressed, that obtained a more widespread circulation among the lay public by Clifford's posthumous book. The section dealing with space was prepared by Karl Pearson, although it was fully based on Clifford's ideas, and it discussed the difficulties inherent in determining the nature of the space that we inhabit. A consequence of non-Euclidean space is that rigid bodies do not necessarily retain their shape when moved, which could be seen as supporting the belief that space is truly Euclidean. Clifford dealt with that point:

> Yet it must be noted that, because a solid figure *appears* to us to retain the same shape when it is moved about in that portion of space with which we are acquainted, it does not follow that the figure *really* does retain its shape. The changes of shape may be either imperceptible for those distances through which we are able to move the figures, or if they do take place we may attribute them to 'physical causes' – to heat, light, or magnetism – which may possibly be mere names for variations in the curvature of our space.

These speculations led to his questioning the existence of the luminiferous aether:

> The most notable physical quantities which vary with position and time are heat, light and electro-magnetism. It is these that

we ought to peculiarly consider when seeking for any physical changes, which may be due to changes in the curvature of space. If we suppose the boundary of any arbitrary figure in space to be distorted by the variation of space-curvature, there would, by analogy from one and two dimensions, be no change in the volume of the figure arising from such distortion. Further, if we *assume* as an axiom that space resists curvature with a resistance proportional to the change, we find that waves of 'space-displacement' are precisely similar to those of the elastic medium [the luminiferous aether] which we suppose to propagate light and heat. We also find that 'space-twist' is a quantity exactly corresponding to magnetic induction, and satisfying relations similar to those which hold for the magnetic field. It is a question whether physicists might not find it simpler to assume that space is capable of a varying curvature, and of a resistance to that variation, than to suppose the existence of a subtle medium pervading an invariable homaloidal space [a space of zero curvature].

Having introduced the hypothesis of curved space, he warned against inferring the universal geometry of space from our particular local experience: not only may our measurements not be sufficiently precise, but also the nature of space may change from place to place, and from time to time. Therefore, he concluded, hypotheses invoking curved space must not be rejected just because 'they may be opposed to the popular dogmatic belief in the universality of certain geometrical axioms – a belief which has arisen from centuries of indiscriminating worship of the genius of Euclid'.[57] In 1882 an anonymous reviewer of Clifford's *Collected Works* encapsulated that view in *The Times*: 'Geometry is, in fact, a physical science; it is a matter of experience and observation. This is the new standpoint.'[58]

Of course, this was not a fully worked-out theory, but it demonstrates not only that Clifford was aware of the implications of non-Euclidean geometry for our understanding of space, time and motion, but also that he was prepared to offer these thoughts in a book that was aimed at the lay public and that achieved a wide readership. He was not a marginal pure mathematician engaging in a pure-mathematical geometrical game for the pleasure of the intellectual challenge; rather, he was offering a radical response to the accumulating problems that the hypothesis of the aether was generating. Furthermore, by using the phrase 'whether physicists might not find it simpler to assume' he was emphasising that, in his view, the function of these hypotheses was not to assert existence, but rather to generate definitions and systems of equations that would constitute an effective mathematical model. Joseph Larmor, a believer in some sort of aether, made a similar point when in 1900 he wrote of Kelvin's model of rotational elasticity:

> the object of a gyrostatic model of the rotational aether is not to represent its actual structure, but to help us to realize that the scheme of mathematical relations which defines its activity is a legitimate conception. Matter may be and likely is a structure in the aether, but certainly aether is not a structure made of matter. This introduction of a suprasensual aethereal medium, which is not the same as matter, may of course be described as leaving reality behind us: and so in fact may every result of thought be described which is more than a record or comparison of sensations.[59]

The use of the aether to explain physical phenomena enjoyed a long life, and persisted well after Einstein published his fully worked-out general theory of relativity in 1915.

The foregoing accounts have shown how well-entrenched was the British tradition in which there was no clear line separating mathematics from its implementation; rather, there was a network of mutual influences between observation, experiment and mathematical theory such that each informed and advanced the others. The holders of Newton's old chair at Cambridge, the Lucasian, most exemplified this tradition, and for most of the Victorian era its occupant was George Gabriel Stokes. The son of a clergyman, he was born in Ireland in 1819 and matriculated at Pembroke College in 1837. After emerging from the Mathematical Tripos as senior wrangler and first Smith's prizeman he was awarded a fellowship before taking up his chair in 1849; he occupied it for 54 years until he died in 1903 at the age of 83. The importance of mathematical physics at Cambridge in the Victorian era can be judged from the Smith's Prize examination paper that he set in 1854; this was the paper taken by Maxwell, and it contained 17 questions, of which 9 appear to partake as much of physics as of mathematics: for example, question 16 is 'Explain the different modes of determining the Mass of the Moon'. Question 8 is particularly interesting because it contains the first public statement of what has become known as Stokes's theorem, although this attribution is misleading because the result had been sent to him by William Thomson as an extension of an existing theorem. Unusually, the question was to have a more significant role in history than that of determining who were to be the Smith's prizemen, because Maxwell took notice of Stokes's theorem and made good use of it when he came to develop his electromagnetic theory; and in a modernised form it is familiar to all mathematicians and physicists today, with a very wide range of applications.

During Stokes's long tenure of the Lucasian chair he made outstanding contributions to mathematical physics, and many phenomena, particularly in fluid dynamics, are named after him. He had an unusually high public profile: for more than 30 years he was a secretary of the Royal Society, and was its president from 1885 to 1890; from 1887 to 1892 he was a member of parliament for the university (it returned two); and in 1889 he was created a baronet. The recipient of many honorary degrees and awards, he also gained public attention by assisting government enquiries into two disastrous bridge collapses – of the Dee Bridge in 1847, which killed five people, and of the first Tay Bridge in 1879, in which at least 59 died; after the latter collapse he was appointed to a commission set up by the Board of Trade to consider the question of wind loading on bridges.

However, it was William Thomson who was perhaps the most complete applied mathematician of the age, and most consistently in the public eye. He was born in Belfast in 1824, but the family moved to Glasgow in 1832 when his father was appointed professor of mathematics at Glasgow College. William was a precocious child, and at the age of 10 started attending his father's elementary classes, where he developed such an affinity with science and technology that at the age of 16 he had his first paper published in the *Cambridge Mathematical Journal*, under the pseudonym 'P Q R'.[60] In the same year, 1841, he began his studies at Cambridge, graduating in 1845 as second wrangler and first Smith's prizeman. During his undergraduate career he became attracted to Faraday's theories of electromagnetism, and it was in this department of science that his first substantial mathematical investigations were carried out.

Then in 1846 what appeared to be a promising but conventional career at Cambridge was cut short when he was appointed to the chair of natural philosophy at Glasgow, where he remained until his retirement, despite receiving offers of more elevated positions elsewhere. At first he did not have a high public profile, for he became engaged in the contemporary debate concerning the nature of heat and work, and the connection between them. The standard theory was that heat was a substance – caloric – that flowed from a hotter body to a colder, but this view was being challenged by the Manchester brewer James Prescott Joule, whose precise experiments suggested that heat and work were mutually convertible, a finding that was not consistent with the caloric theory. Thomson began as a believer in caloric, but eventually concluded that Joule was right, and that energy could be used as a unifying concept in terms of which both heat and work could be expressed.[61] This was of considerable interest to many scientists and engineers, but not of such a nature as to put him firmly in the public eye.

That changed in the mid 1850s when he became involved with the transatlantic cable that the Atlantic Telegraph Company was proposing to lay from the west coast of Ireland to Nova Scotia, via Newfoundland. Much of the theory up until then had been developed by Faraday, and a short cable under the English Channel had been laid successfully, but to lay a cable connecting Britain and America brought problems of a completely different order. In particular, the great distance raised an important question, namely the extent to which the length of the cable, and the material of which it was made, affected the efficiency with which data could be transmitted; if the speed was too slow, or the signal too weak, then the economic viability of the whole enterprise was in doubt. Disagreement on this issue brought Thomson into direct conflict with the company's chief electrician, the intriguingly named Edward Orange Wildman Whitehouse – a medical man by profession – under whose supervision the project was already well under way.

Notwithstanding this difference of opinion, by the end of 1856 the board of directors had been sufficiently impressed by Thomson to offer him a directorship. From the start he had had reservations about Whitehouse's ability to bring the project to a successful conclusion, and had made many suggestions for improvement, but these were not taken up. Nevertheless, in August 1857 he was on board *HMS Agamemnon* for the first attempt to lay a cable, which had to be abandoned when the cable broke. In 1858 a second attempt, again with Thomson on board, was eventually successful, but Whitehouse then wrecked the cable by applying a load far greater than that for which it was designed, in a vain attempt to achieve the efficiency in transmission that he had promised. This disaster could have been avoided, for Thomson had already developed a greatly improved mirror galvanometer, an instrument in which a mirror, on which small magnets have been secured, undergoes a minute deflection when in the presence of the electromagnetic field generated by a very faint electrical impulse such as was received at the far end of the cable; a spot of light is then reflected by the mirror on to a screen, which has the effect of amplifying and making easily visible the otherwise imperceptible deflection. Although the mirror galvanometer was not Thomson's invention, his improvements, which he patented, meant that it could detect much weaker signals than could any other design. He had hoped that he would be able to use it in 1858, but was not given the opportunity; if his instrument *had* been used, it is likely that Whitehouse would never have needed to overload the cable.[62]

After this debacle Whitehouse was dismissed, and Thomson took on a much more prominent role. In the two years that he had been involved with the project he had shown himself to be not just a mathematician but also a practical engineer, and so he was appointed to a committee set up by the Board of Trade to enquire into the events of 1858; their report in 1863 placed most of the blame on Whitehouse. Plans were then made for another cable, and in July 1865 Thomson found himself on board Isambard Kingdom Brunel's *SS Great Eastern*, then the largest ship ever built, which had been converted for cable-laying duties. Unfortunately the cable was lost and the attempt abandoned, but the next year a new cable was successfully laid and the lost cable recovered; so with the aid of some of Thomson's instruments, telegraph messages could at last be exchanged between Britain and America. From its inception the whole project had been extremely newsworthy, not least because at various times capital had been raised from the public, and therefore many investors had a personal interest in the project's fortunes. Its eventual success was widely celebrated as a national triumph, and for this Thomson received much of the credit; he was duly knighted in November 1866.

The story of the transatlantic cable illustrates well the way in which the work of an applied mathematician impinged on everyday lives in a manner that the work of a pure mathematician could not; and Thomson combined the mathematical insight of Maxwell with the inventiveness of Faraday, the practical skills of an engineer, and the commercial acumen of an entrepreneur. He was clearly the right man in the right place at the right time, because during the following decade he was involved with several more cable-laying projects, and became particularly drawn to problems concerning navigation, which led to what was perhaps his most famous invention – an improved mariners' compass. The large quantities of iron that were increasingly to be found in modern ships caused magnetic compasses to show major errors, and Thomson's system was successful in minimising these; the patents that he took out, and which he jealously protected by legal action, provided him with a secure income for the rest of his life. Furthermore, during the second half of the nineteenth century electricity moved from being the subject only of scientific interest to being a commercial source of power, and Thomson was foremost in using his talents to devise some of the new instruments for measurement and calibration that this expansion of activity required.

Like Faraday, Maxwell and Stokes, Thomson had a strong religious bent, but also like them he did not allow religion to interfere with his science, and in collaboration with Peter Guthrie Tait wrote a *Treatise on Natural Philosophy* that remained a leading textbook until well into the twentieth century. However, a topic of the greatest public interest that could have caused problems for him had his Christianity been of a fundamentalist nature was the age of the Earth, on which he worked intermittently throughout his life. In the seventeenth century James Ussher, the Archbishop of Armagh, had calculated from biblical chronology that God created the world in 4004 BCE, but during the latter half of the eighteenth century belief in the correctness of this date, or anything like it, had started to decline in the face of mounting geological evidence to the contrary. By the time that Thomson began his investigations some contemporary geologists were arguing that the Earth was almost unimaginably old, but their conclusions relied on an assumption with which he strongly disagreed, and which he explained thus:

> For eighteen years it has pressed on my mind, that essential principles of Thermo-dynamics have been overlooked by those geologists who uncompromisingly oppose all paroxysmal hypotheses, and maintain that not only do we have examples now before us, on the earth, of all the different actions by which its crust has been modified in geological history, but that these actions have never, or have not on the whole, been more violent in past time than they are at present.[63]

Thomson believed that geologists were unjustifiably excluding from consideration the possibility of natural events that could have brought about the present geological state of the Earth much more quickly. His preferred approach was to make calculations of the effect of tidal friction on the Earth's rotation, of the rate at which the Earth had been cooling, and of the time for which the Sun could have been yielding heat by gravitational collapse, which was then the only known means.[64] In 1861 he addressed the question of the age of the Sun in a contribution to the popular general-interest *Macmillan's Magazine*, pouring scorn on geologists' calculations: 'What then are we to think of such geological estimates as 300,000,000 years for the "denudation of the Weald?"' – this having been suggested by Charles Darwin. He concluded:

> It seems, therefore, on the whole most probable that the sun has not illuminated the earth for 100,000,000 years, and almost certain that he has not done so for 500,000,000 years. As for the future, we may say, with equal certainty, that inhabitants of the earth cannot continue to enjoy the light and heat essential to their life, for many million years longer, unless sources now unknown to us are prepared in the great storehouse of creation.[65]

As this shows, Thomson's probable upper limit for the age of the Sun was one-third of the time that had been proposed merely for the 'denudation of the Weald'. He continued to refine the calculations throughout his life, but as the timescale derived from geological evidence grew longer and more strongly supported, so the upper limit of his estimates reduced. In his final years the discovery of radioactivity as a source of energy invalidated those of his calculations that dealt with the cooling Earth, but did not affect those concerning the age of the Sun, for it was not until long after his death that thermonuclear fusion was understood as being the source of the Sun's heat. Consequently he never abandoned his comparatively brief timescale, because he considered that the duration of the Sun's gravitational collapse was more significant than the rate at which the Earth was cooling. Despite his conclusions being so much in error, his prescient rider to the effect that they would need revision if some previously unsuspected source of energy was discovered means that posterity has been quite kind in respect of his judgement.

During his long and active life Thomson received a great many public honours both at home and abroad, and served as president of the

Royal Society from 1890 to 1895. He was also president of the London Mathematical Society for two years from 1898 – he had joined in 1867, but rarely attended meetings due to his residence in Scotland. In 1892 he was ennobled as Baron Kelvin of Largs; consequently the base unit of the absolute scale of temperature that he developed is referred to as the kelvin, and this name is attached to many developments in applied mathematics, theoretical physics and engineering. Ten years later, in 1902, he was appointed to the Privy Council. When he died in 1907 the obituary in *The Times*, which took up almost a whole page, referred to him as 'the most distinguished British man of science',[66] and a few days later he was interred in Westminster Abbey, close to the grave of Isaac Newton; nearby is a memorial window donated in 1913 by engineers in Britain and America.[67] Yet although he was such a famous and high-profile figure, his achievements were not such as to merit a Nobel Prize, for which he was nominated seven times by eminent physicists.[68]

The prize was first awarded in 1901, and such was the strength of Britain in applied mathematics and physics during the second half of the nineteenth century that two Britons were to be amongst the early recipients for their work in these fields – Lord Rayleigh in 1904, and Sir Joseph John Thomson in 1906. Most of the work for which J J Thomson became famous falls at the very end of the Victorian period, and so he will not be considered further here, but Rayleigh was one of the most noteworthy scientists in the second half of the nineteenth century. He was born John William Strutt in 1842 to parents who were members of the minor nobility, and who had substantial agricultural land holdings in Essex that made them financially independent; he inherited the title and estates from his father, the second Baron Rayleigh, in 1873. Strutt did not enjoy good health as a boy, so his education was a mixture of schooling and private tutoring, after which he decided to study for the Mathematical Tripos at Trinity College, Cambridge. After matriculating he enlisted the services of Edward Routh, probably the most successful of all the Cambridge mathematical coaches, and emerged in 1865 as senior wrangler and first Smith's prizeman. The following year he was elected a Fellow of Trinity, and shortly afterwards set up a private laboratory at Terling Place, the family seat in Essex, where he began the investigations that were to occupy him for the rest of his life.

Even more than Kelvin, Rayleigh straddled the two worlds of the theoretician and the experimentalist, a circumstance that is reflected in his publications, which numbered more than 400. As might be expected of a senior wrangler, they included papers on mathematical topics such as Laplace, Legendre and Bessel functions, but also on many of those aspects of physics that most preoccupied the Victorians. He was an active supporter of the London Mathematical Society, gaining membership in 1871, making a gift of £1,000 when insolvency loomed in 1874, and acting as president from 1876 to 1878. Up to the turn of the century he published in the Society's *Proceedings* 23 papers, totalling about 174 pages, which were mostly on various aspects of applied mathematics. This was only a small proportion of his total output, but contributions from such a distinguished figure gave valuable support to the Society's efforts to appear more relevant to applied mathematicians.

In his early days he was concerned with electromagnetic theory; like so many he had been inspired by Maxwell, and he explained the blueness of the sky as being an electromagnetic effect that occurs when light is incident on particles that are much smaller than the light's wavelength, although this is not in accord with modern theories.[69] Phenomena that are characteristically manifested by wave motions particularly interested him; he wrote many papers on the mathematical treatment of waves, and during 1877 and 1878 published a classic two-volume treatise on *The Theory of Sound*. His career then took an unexpected turn when Maxwell died in 1879 and the chair of experimental physics at Cambridge became vacant. Rayleigh was the most obvious candidate for the appointment, and accepted, partly at least because the agricultural crisis of the 1870s had reduced his income.

After six very successful years as both teacher and researcher he resigned and returned to his laboratory at Terling Place, from where most of his subsequent work was done, although he served as professor of natural philosophy at the Royal Institution from 1887 to 1905. In 1890 he was the third recipient of the De Morgan medal from the LMS, and in addition to the Nobel Prize he received many honours and awards both in Britain and overseas. He was also active in public life, being Lord Lieutenant of Essex from 1892 to 1901, president of the Royal Society from 1905 until 1908, and Chancellor of Cambridge University from 1908 until his death in 1919. Like Kelvin, he was appointed a privy councillor, and (also like Kelvin) was among the first batch of notables to receive the Order of Merit when it was instituted in 1902. All these honours were in recognition of characteristics that *The Times* summed up in in its obituary: 'Lord Rayleigh was marked out by the scope, the amount, and the quality of his scientific investigations as one of the foremost British mathematical physicists of his generation. A leading characteristic of his work was its meticulous accuracy and finish – qualities as apparent in the form of his mathematical analysis, as in his spoken and written prose.'[70] It was this devotion to 'meticulous accuracy' that led to some of his most memorable results, including the standardisation of the ohm – the unit of electrical resistance – and his Nobel-Prize-winning discovery of the inert element argon, after more than 20 years' investigation into the densities of gases.

In addition to his membership of the House of Lords and his scientific achievements, Rayleigh's public profile was further enhanced by his being extremely well-connected through his wife, Evelyn Balfour. Her brother Arthur was conservative prime minister from 1902 to 1905, and her sister Eleanor – who acted for a time as Rayleigh's laboratory assistant – married Henry Sidgwick, one of the foremost philosophers of the day. Sidgwick was a founder of the Society for Psychical Research, which aimed to investigate psychic phenomena with scientific rigour, and given the fascination that action at a distance and unseen forces had for the Victorians, it is unsurprising that Rayleigh was also drawn towards the subject, as were several other prominent Victorian scientists such as Oliver Lodge and William Crookes. Rayleigh was an early member of the society, and became president in 1919, the year of his death.

The final mathematician in the roll call of those who achieved widespread public recognition for their achievements in the applied sciences is George Biddell Airy. He was born in 1801 in Northumberland, and after a somewhat peripatetic childhood went up to Trinity College, Cambridge, to study mathematics; there he was an outstanding student, emerging in 1823 as senior wrangler and first Smith's prizeman. In 1826, at the age of 25, he was appointed Lucasian Professor of Mathematics, but 14 months later transferred to the Plumian chair of astronomy and experimental philosophy, which was much better paid; he was always careful about money. During his time in this post he also acted as the director of the Cambridge Observatory, where he became noted for the combination of hard work and administrative efficiency that was to characterise his career. He first came to wider notice when, in 1830, he calculated the parameters for the Airy ellipsoid – an approximation to the shape of the Earth obtained by rotating an ellipse around its shorter axis – upon which the reference system of latitude and longitude coordinates could be superimposed. For greatest accuracy, different geographical areas need to be approximated by sections of slightly different *reference ellipsoids*, and Airy ensured that his ellipsoid provided a particularly good fit for Great Britain, so that it could be adopted by the Ordnance Survey as the basis of their map-making. Shortly afterwards, and at the request of the newly established British Association for the Advancement of Science, he prepared a lengthy report on the progress of nineteenth-century astronomy that was published in 1833.[71]

However, Airy's worldwide recognition began with his appointment in 1835 as Astronomer Royal, based at the Greenwich Observatory; he was to remain in that role for 46 years. On moving to Greenwich his personal qualities were to stand him in very good stead, for the Observatory was in a poor state as regards its organisation and equipment, a situation that he tackled with great energy. Most of the work of the Observatory consisted of making accurate measurements of astronomical phenomena, processing the data with the aid of human computers, and publishing tables based on the results. There was also a great quantity of historical observational data that had never been properly processed, so Airy's very formal, structured approach to organising his own work and the work of others was ideally suited to putting the Observatory on a sound footing.

In 1851 the first observations were made with the Great Transit Circle, which was destined to become one of the world's most famous astronomical instruments, and had been built to Airy's specification; accordingly, it was better known as the Airy Transit Circle. Its centrepiece was a telescope permanently orientated in a north–south direction (so that it could move only vertically), because its purpose was to record both the time at which a star was observed to cross the meridian, and the altitude of the star above the horizon at that moment, data that were then used to regulate Greenwich Mean Time. The concept of a transit circle was an old one, but Airy's design allowed unprecedented accuracy of observation, to the extent that the position of a star above or below the celestial equator could be calculated to within $\frac{1}{360,000}$ of a degree.[72] Greenwich Mean Time was the basis of international timekeeping, so the clocks of the world were ultimately regulated by Airy's instrument, a circumstance that was formally recognised when at the International Meridian Conference of 1884 it was agreed that the north–south vertical plane passing through the centre of the eyepiece of the Great

Circle's telescope should define the prime meridian, or zero degrees of longitude.[h]

There were many other instruments, less famous than the Great Transit Circle, that Airy introduced into the Observatory, and some of these he designed himself, adopting rather unusual principles. The astronomer Simon Newcomb, one of the most substantial figures in American science during the nineteenth century, visited Airy at Greenwich during the winter of 1870, and later reported:

> Before his time the trained astronomer worked with instruments of very delicate construction, so that skill in handling them was one of the requisites of an observer. Airy made them in the likeness of heavy machinery, which could suffer no injury from a blow of the head of a careless observer. Strong and simple, they rarely got out of order. It is said that an assistant who showed a visiting astronomer the transit circle sometimes hit it a good slap to show how solid it was ...[73]

Airy also deployed his considerable mathematical skills in devising new methods that allowed the human computers to process more rapidly the large quantity of data that the Observatory was producing. Newcomb saw at first-hand how Airy had transformed the computers' way of working:

> Under the new system they needed to understand only the four rules of arithmetic; indeed, so far as possible Airy arranged his calculations so that subtraction and division were rarely required. His boys had little more to do than add and multiply. Thus, so far as the doing of work was concerned, he introduced the same sort of improvement that our times have witnessed in great manufacturing establishments, where labor is so organized that unskilled men bring about results that formerly demanded a high grade of technical ability. He introduced production on a large scale into astronomy.[74]

---

[h] A new coordinate system was introduced in 1984 to deal with the requirements of satellite navigation; consequently, all longitudes were revised. At Greenwich this meant that the prime meridian was relocated approximately 102 metres east of Airy's Transit Circle, as puzzled visitors with GPS devices can confirm. See Stephen Malys et al., 'Why the Greenwich Meridian Moved', *Journal of Geodesy*, 89, 12 (2015), 1263–1272.

Airy's extraordinary ability to get through business – allied with his considerable intellect, wide-ranging interests and practical cast of mind – ensured that, in addition to his regular duties as Astronomer Royal, he was called upon by the government to advise and report on a wide range of matters, many of which had very little relevance to astronomy.

As might be expected of a senior wrangler, Airy also involved himself with more theoretical issues, particularly if they had implications for the measurement of astronomical phenomena. In 1867 the German astronomer Ernst Klinkerfues published certain observations that he had made concerning aberration, together with a calculation of what the observations should have been if the undulatory (or wave) theory of light was correct; but the observations and the calculation did not agree. By 1871 Airy had become much exercised by this disagreement, and believed that it was 'a result of great physical importance, not only affecting the computation of the velocity of light, but also influencing the whole treatment of the Undulatory Theory of Light'. He therefore determined to test Klinkerfues's findings with an experiment in which a water-filled telescope was used to observe the star γ-Draconis; this was the star with which Bradley had discovered aberration, and it was particularly suitable for such observations because its zenith was almost directly overhead at Greenwich. Airy concluded – one senses, to his relief – that Klinkerfues's findings were wrong, and that the wave theory was safe; but the results of the experiment were difficult to square with the existence of the aether, which only added to the general uncertainty on this question. Consequently, Airy's experiment was regarded as being of great significance, and is often considered to be a milestone on the road to Einstein's special theory of relativity.[75]

Airy's name was given even more exposure to the general public by his book *Popular Astronomy*, which by 1881 had reached its tenth edition. Like Kelvin and Rayleigh, he received numerous national and international awards for his work, including two Gold Medals from the Royal Astronomical Society, of which he four times served as president. He was also president of the Royal Society for two years from 1871, and in 1872 accepted a knighthood for his public services. His reputation was not entirely untarnished, however: he was much criticised for apparent inaction that resulted in Continental astronomers, rather than British, being credited with the discovery of the planet Neptune, the existence and position of which had been inferred, using Newtonian gravitational theory, from anomalies in the orbit of Uranus; and after the collapse of the first Tay Bridge in 1869 there was called into question the advice on wind speeds and pressures that he had given to Thomas Bouch, the design engineer who was held largely responsible for the disaster. However, Airy's work as Astronomer Royal ensured that for much of the Victorian era he was one of Britain's best-known scientific figures, because under his leadership the Greenwich Observatory had gained world pre-eminence in its field, thanks to the quantity, authority and precision of its published data.

The Victorians saw great changes in the applied sciences, and this is particularly so as regards the theories of light, electricity and magnetism, together with their concomitant, the aether. In this chapter we have used that circumstance as a theme with which to illustrate how mathematicians who were regarded by their peers as being some of the finest practitioners of the age – Maxwell, Rayleigh, Thomson, Stokes, Clifford and Airy – could achieve a high public profile, notwithstanding the occult nature of their primary discipline, mathematics. They were drawn from a very select band, namely those who had emerged from the Mathematical Tripos at Cambridge as senior or second wranglers, and were therefore skilled exponents of the traditional problem-solving style of mathematics, rooted in natural philosophy, that had reigned there since the time of Newton. Although some of these mathematicians had undertaken preliminary studies or taught elsewhere, Cambridge was clearly the nation's mathematical powerhouse, and was seen to be so because of the contributions that some of its most distinguished wranglers had made to public life, contributions that had been rewarded in particular by the ennobling of William Thomson (previously knighted), the baronetcy bestowed on Stokes, and the knighthood awarded to Airy. In contrast to the work of pure mathematicians such as Cayley, Sylvester and Glaisher, the developments in mathematical physics engaged the public at large not only through practical applications such as the telegraph, electrical power, geodesy and engineering, but also through subjects of wider scientific interest such as astronomy, the nature of space and time, and the age of the Earth. Public understanding of the value of mathematics did not derive from the mathematics itself, but on what it could do for us, and what it could tell us about the world.

The virtue of mathematics, from this point of view, was that when allied to imaginary mechanical models it yielded mathematical models with explanatory and predictive power. This predictive power was not necessarily limited to the realm of quantitative observations of phenomena already known; from a mathematical model could be inferred (but not proved) the existence of things hitherto unsuspected, such as the planet Neptune, and electromagnetic waves. This was all 'value added' to mathematical activity, and was conveyed to the public through news reports, articles in magazines and journals aimed at a general audience, and popular books and lectures. A notable feature of Victorian culture was an enthusiasm for science, and we saw at the beginning of Chapter 3 how this resulted in a blossoming of popular scientific societies. Only one of these – the London Mathematical Society – was overtly mathematical, but many others dealt with matters that were elucidated by mathematics, and thereby helped to spread appreciation of its power and usefulness.

It was against this background that the new pure mathematicians were struggling to make an impact, both on their mathematical colleagues and on the public; yet utility, contributing to the prosperity of the Empire, increasing our knowledge of the world, furthering our understanding of the nature of the Universe – these were things that they could not offer, and did not wish to offer. The efforts that they made to surmount this apparently insurmountable difficulty, in an ultimately successful effort to establish and then consolidate their newfound independence, will be the subject of our final chapters.

Notes and References

René Descartes gave the most complete account of his system of physics in Renate Descartes, *Principia Philosophiae* (Amsterdam: Elzevir, 1644); it was written in Latin, but with his approval an expanded edition was published in French in 1647. There have been many republications, editions and translations since then; however, they sometimes omit those parts of the book that offer detailed explanations of natural phenomena, for that is not the focus of interest for most modern readers. The translation of the whole work into English by Valentine Rodger Miller and Reese P Miller (Dordrecht: Reidel, 1983) has been used in writing this chapter.

1. Descartes, *Principles of Natural Philosophy*, II, p. 16.
2. Descartes, *Principles*, IV, p. 195.
3. See Hooke's paper delivered to the Royal Society in 1672, in Thomas Birch, *The History of the Royal Society of London for Improving of Natural Knowledge ...*, 3 (London: The Royal Society, 1756), pp. 10–15.
4. Newton's letter to the Royal Society setting out the results of his experiments and his theory of light and colours was published in *Philosophical Transactions*, 80 (19 February 1671/2), 3075–3087; his response to Hooke's paper is in *Philosophical Transactions*, 88 (18 November 1672), 5084–5103.

5. Letter from Thomas Young to Edward Whitaker Gray dated 8 July 1799: 'Outlines of Experiments and Inquiries Respecting Sound and Light', *Philosophical Transactions of the Royal Society of London*, 90 (1800), 106–150, at 126.
6. E T Whittaker, *A History of the Theories of Aether and Electricity from the Age of Descartes to the Close of the Nineteenth Century* (London: Longmans, Green, 1910), pp. 115–116.
7. Letter from Thomas Young to François Arago dated 12 January 1817, in George Peacock (ed.), *Miscellaneous Works of the late Thomas Young*, 1 (London: John Murray, 1855), p. 383.
8. Joseph Larmor, *Aether and Matter: A Development of the Dynamical Relations of the Aether to Material Systems on the basis of the Atomic Constitution of Matter, including a Discussion of the Influence of the Earth's Motion on Optical Phenomena* (Cambridge: Cambridge University Press, 1900), p. 11.
9. James MacCullagh, 'An Essay towards a Dynamical Theory of Crystalline Reflexion and Refraction', *The Transactions of the Royal Irish Academy*, 21 (1846), 17–50.
10. Whittaker, *History*, p. 269.
11. Michael Faraday, 'On some new Electro-Magnetical Motions, and on the Theory of Magnetism', *Quarterly Journal of Science, Literature and the Arts*, XII (1822), 74–96.
12. Michael Faraday, 'Historical Statement respecting Electro-Magnetic Rotation', *Quarterly Journal of Science, Literature and the Arts*, XV (1823), 288–292.
13. Michael Faraday, 'Experimental Researches in Electricity', *Philosophical Transactions of the Royal Society of London*, 122 (1832), 125–162, at 149.
14. Michael Faraday, 'On lines of Magnetic Force; their definite Character; and their Distribution within a Magnet and through Space', *Philosophical Transactions of the Royal Society of London*, 142 (1852), 25–56; he had been using the term 'lines of force' for many years, but with a lack of clarity that he aimed to correct in this paper.
15. Michael Faraday, 'Thoughts on Ray-vibrations' [Letter to Richard Phillips dated 15 April 1846], *London, Edinburgh and Dublin Philosophical Magazine and Journal of Science*, 3, 28, 188 (May 1846), 345–350.
16. J Clerk Maxwell, *Theory of Heat* (London: Longmans, Green, 1871), pp. 310–311.
17. All quotations from Maxwell, *Theory of Heat*, pp. 308–309.
18. J Clerk Maxwell, 'On Faraday's Lines of Force', *Transactions of the Cambridge Philosophical Society*, 10, 1 (1864), 27–83; the paper was read on 10 December 1855 and 11 February 1856. P G Tait, a friend of Maxwell's since their schooldays, recalled seeing the manuscript of most of the paper in 1853: 'James Clerk Maxwell', *Proceedings of the Royal Society of Edinburgh*, 10 (1878–1880), 331–339, at 334.
19. Tait, 'James Clerk Maxwell', 334.
20. Albert Einstein (tr. & ed. Paul Arthur Schilpp), *Autobiographical Notes* (Open Court: La Salle, IL, 1979), p. 31; originally published 1949.
21. J Clerk Maxwell, 'A Dynamical Theory of the Electromagnetic Field', *Philosophical Transactions of the Royal Society of London*, 155 (1865), 459–512, at 460.
22. J C Maxwell, 'On Physical Lines of Force, III: The Theory of Molecular Vortices', *The London, Edinburgh and Dublin Philosophical Magazine and Journal of Science*, 4, 23, 151 (January 1862), 12–24, at 14.
23. Maxwell, 'On Physical Lines of Force, III', 19.
24. Maxwell, 'On Physical Lines of Force, III', 21–22.
25. Maxwell, 'A Dynamical Theory', 499.
26. Basil Mahon, *The Man who Changed Everything: The Life of James Clerk Maxwell* (Chichester: Wiley, 2003), pp. 176–179.
27. Einstein, *Autobiographical Notes*, p. 59.

28. William Kingdon Clifford, *Seeing and Thinking* (London: Macmillan, 1879), pp. 38–39.
29. G G Stokes, 'On the Constitution of the Luminiferous Aether', *The London, Edinburgh and Dublin Philosophical Magazine and Journal of Science*, 3, 32, 216 (1848), 343–349, at 347–348.
30. Larmor, *Aether and Matter*, p. 13.
31. Thomas Young, 'The Bakerian Lecture. Experiments and Calculations relative to Physical Optics', *Philosophical Transactions of the Royal Society of London*, 94 (1804), 1–16, at 12–13.
32. Larmor, *Aether and Matter*, pp. 16–17.
33. Letter from J Clerk Maxwell to D P Todd dated 19 March 1879, *Nature*, XXI (1879–80), 315. Maxwell had previously made the same point in an article on 'Ether' in *The Encyclopædia Britannica*, ninth edition, VIII (New York: Scribner's, 1878), p. 570.
34. Albert A Michelson, 'On the Relative Motion of the Earth and the Luminiferous Ether', *The American Journal of Science*, 3, XXII, 128 (August 1881), 120–129.
35. Albert A Michelson and Edward W Morley, 'On the Relative Motion of the Earth and the Luminiferous Ether', *The American Journal of Science*, 3, XXXIV, 203 (November 1887), 333–345, at 341.
36. See Letter to the Editor from G F Fitz Gerald: 'The Ether and the Earth's Atmosphere', *Science*, XIII, 328 (May 1889), 390.
37. See the Nobel Prize archive available online at http://www.nobelprize.org/nomination/archive/show_people.php?id=3972
38. Oliver Heaviside, *Electromagnetic Theory*, 2 ('The Electrician': London: 1899), p. 10.
39. Letter from Heaviside to Heinrich Hertz dated 13 July 1889, quoted in Paul J Nahin, *Oliver Heaviside: The Life, Work and Times of an Electrical Genius of the Victorian Age* (Baltimore, MD: Johns Hopkins, 2002), p. 111; originally published 1987.
40. Oliver Heaviside, *Electromagnetic Theory*, 1 (London: 'The Electrician', 1893), p. 134.
41. Arthur Eddington, 'Joseph Larmor', *Obituary Notices of Fellows of the Royal Society*, 4 (November 1942), 197–207, at 198.
42. From the stenographer's record of lectures that Kelvin gave at Johns Hopkins University, Baltimore, in October 1884. The record is reproduced in its entirety in Robert Kargon and Peter Achinstein (eds.), *Kelvin's Baltimore Lectures and Modern Theoretical Physics: Historical and Philosophical Perspectives* (Cambridge MA: MIT Press, 1987); the quotation is from Lecture XX, p. 206. The word 'get', in this context, appears to be an Americanism, which suggests that the notes may not be verbatim, but rather the stenographer's rendering of what Kelvin said: for this suggestion, see T H R Skyrme, 'The Origins of Skyrmions', in Gerald E Brown (ed.), *Collected Papers, with Commentary, of Tony Hilton Royle Skyrme* (Singapore: World Scientific, 1994), pp. 119–120 and references 2 and 3 on p. 125. However, there is good reason to suppose that the quotation is substantially accurate, because Kelvin later thanked the stenographer 'for the care and fidelity with which he stenographically recorded my lectures': see Lord Kelvin (William Thomson), *Baltimore Lectures on Molecular Dynamics and the Wave Theory of Light* (London: C J Clay, 1904), p. vii. Nevertheless, the confession of his need for a mechanical model does not appear in this volume, which is a substantial reworking of the original lectures.
43. Whittaker, *History*, p. 157.

44. Whittaker, *History*, p. 336.
45. J J Sylvester, 'Address', *Report of the Thirty-ninth Meeting of the British Association for the Advancement of Science* [1869] (London: John Murray, 1870), p. 4 of 'Notices and Abstracts'.
46. Hermann von Helmholtz, 'The Axioms of Geometry', *The Academy*, I, 5 (12 February 1870), 128–131; and Letter to the Editor, 'The Axioms of Geometry', *The Academy*, III, 41 (1 February 1872), 52–53.
47. Bernhard Riemann (tr. William Kingdon Clifford), 'On the Hypotheses which lie at the Bases of Geometry', *Nature*, 8, 183–184 (1 & 8 May 1873), 14–17 & 36–37.
48. Joan Richards, *Mathematical Visions: The Pursuit of Geometry in Victorian England* (San Diego, CA: Academic Press, 1988), p. 74.
49. Robert Ball, 'The Non-Euclidean Geometry', *Hermathena*, 3, 6 (1879), 500–541.
50. [A Square] Edwin Abbott Abbott, *Flatland: A Romance of Many Dimensions* (London: Seeley, 1884).
51. William Kingdon Clifford, *The Common Sense of the Exact Sciences* (London: Kegan Paul, 1885). The book was completed after his death by Karl Pearson, who was responsible for pp. 116–226, i.e. the latter part of chapter III ('Quantity') and the whole of chapter IV ('Position'); see Pearson's preface, pp. v–x.
52. Isaac Todhunter, *The Elements of Euclid, for the Use of Schools and Colleges; comprising the First Six Books and Portions of the Eleventh and Twelfth Books; with Notes, an Appendix, and Exercises* (Cambridge: Macmillan, 1862), p. 6.
53. See, for example, A M Legendre, *Éléments de Géométrie, avec des Notes*, Septième Édition (Paris: Firmin Didot, 1808), in which he reverses Euclid's order by first 'proving' (proposition XX, p. 20) that the internal angles of a triangle sum to two right angles, and then uses this result to 'prove' the fifth postulate (proposition XXIV, p. 25).
54. Bertrand A W Russell, *An Essay on the Foundations of Geometry* (Cambridge: Cambridge University Press, 1897). For a discussion of his early views on geometry, see Nicholas Griffin, *Russell's Idealist Apprenticeship* (Oxford: Clarendon Press, 1991), pp. 135–153.
55. Sylvester, 'Address', 4–5 of 'Notices and Abstracts'.
56. William Kingdon Clifford, 'On the Space-Theory of Matter' [Abstract], *Proceedings of the Cambridge Philosophical Society*, II (1864–1876), 158; the meeting was on 21 February 1870.
57. Quotations from Clifford, *Common Sense*, pp. 223–226.
58. *The Times*, 14 April 1882, p. 4, col. A.
59. Larmor, *Aether and Matter*, p. vi.
60. [P. Q. R., 'A Correspondent'] William Thomson, 'On Fourier's Expansions of Functions in Trigonometrical Series', *Cambridge Mathematical Journal*, II (1841), 258–262.
61. Iwan Rhys Morus, '"A Dynamical Form of Mechanical Effect": Thomson's Thermodynamics', in Raymond Flood, Mark McCartney and Andrew Whitaker (eds.), *Kelvin: Life, Labours and Legacy* (Oxford: Oxford University Press, 2008), pp. 122–123.
62. Charles Bright, *The Story of the Atlantic Cable* (London: Newnes, 1903); Bright was the son of Sir Charles Bright, the engineer responsible for laying the cables.
63. William Thomson, 'On the Secular Cooling of the Earth', *Transactions of the Royal Society of Edinburgh*, XXIII (1864), 157–169, at 157; the paper was read on 28 April 1862.

64. William Thomson, 'On Geological Time', in Sir William Thomson (Baron Kelvin), *Popular Lectures and Addresses*, 2 (London: Macmillan, 1894), pp. 10–72; this address was delivered before the Geological Society of Glasgow, 27 February 1868.
65. William Thomson, 'On the Age of the Sun's Heat', *Macmillan's Magazine*, V (1861–1862), 388–393.
66. *The Times*, 18 December 1907, p. 8.
67. See www.westminster-abbey.org/abbey-commemorations/commemorations/wil liam-thomson-lord-kelvin/#i14245
68. See www.nobelprize.org/nomination/archive/show_people.php?id=9249
69. Barry R Masters, 'Lord Rayleigh: A Scientific Life', *Optics and Photonics News*, 20, 6 (June 2009): www.osa-opn.org/home/articles/volume_20/issue_6/features/lord_rayleigh_a_scientific_life/
70. *The Times*, 2 July 1919, p. 19.
71. G B Airy, 'Report on the Progress of Astronomy in the present Century', *Report of the First and Second Meetings of the British Association for the Advancement of Science* (London: John Murray, 1833), pp. 125–189.
72. The accuracy quoted by the Royal Observatory, Greenwich, is 0.01 seconds of arc: www.rmg.co.uk/discover/explore/airys-transit-circle-and-dawn-universal-day
73. Simon Newcomb, *The Reminiscences of an Astronomer* (Boston, MA: Houghton, Mifflin, 1903), pp. 288–289.
74. Newcomb, *Reminiscences*, p. 288.
75. George Biddell Airy, 'On a Supposed Alteration in the Amount of Astronomical Aberration of Light, produced by the Passage of the Light through a Considerable Thickness of Refracting Medium', *Proceedings of the Royal Society of London*, 20 (1871–1872), 35–39, at 37.

6 APOLOGIAS FOR PURE MATHEMATICIANS

The extent of the public and academic reputations that were enjoyed during the Victorian era by British applied mathematicians explains why Glaisher was so envious of 'the Cambridge school of mathematical physics', whose members were known the world over as having made some of the century's most valuable contributions to science and society. In this chapter we consider whether, against such a backdrop, pure mathematicians were able to deflect criticism, and fashion for themselves a plausible justification for indulging in pure, rather than applied, mathematics.

Allied to this question, and one that needs to be dealt with, is that of professionalisation. This social phenomenon, one of the most significant in Victorian Britain, was closely connected with the rise of the middle class, and amongst historians of science 'the professionalisation of the sciences' is ubiquitous in the literature. If professionalisation had also embraced pure mathematics, as some writers have suggested,[1] then it might have offered pure mathematicians the means of securing for themselves the status that was enjoyed by professions of undoubted public usefulness. To determine the matter, it will be helpful to define two varieties of professionalisation, which we may call social and academic: in the former, there was a requirement that a new profession had to maintain certain mutually beneficent relations with society, whereas the latter can be seen as the beginning of the cult of the 'expert', in which the right to make authoritative pronouncements and judgements in a particular discipline is granted only by virtue of some formal connection with a university department or other relevant institution.

The origin of the professions can be found in the medieval universities, which followed a similar pattern throughout Europe and comprised several faculties, the highest being those of law, theology and medicine. A 'profession', which originally was a vow made when entering a religious order, later widened in meaning to include an exposition of the fruits of these three faculties, and whoever made such a profession was said to be a 'professor' who 'professed' his learning. Hence, by a natural adaptation, 'profession' came to be a collective term for all the 'professors' of one of the three faculties, and so from early times the word had elitist connotations.

In Britain, the concept of a profession came to be intimately connected with that of the gentleman. This personage, who enjoyed his status in virtue of his birth, was by definition not a nobleman but otherwise had sufficient resources to follow his own inclinations, beholden to no one for his livelihood. Such freedom was essential, and a gentleman's word was his bond because he was subject to no pressure that could divert him from the path of absolute rectitude. This was the aspect of gentlemanliness that so appealed to the founders of the Royal Society, and Robert Boyle wrote that a gentleman 'has little else to do, but to conserve by his Actions, that Esteem that his Birth has given which makes Men presume to have merit, till his own Indiscretion declare the Contrary'.[2] Boyle conducted himself according to his own prescriptions, so when in 1673 one of his experimental observations was queried by Gottfried Leibniz, Henry Oldenburg, the secretary of the Royal Society, was able to reply simply: 'Boyle says that that experiment of his ... was related by himself in good faith'.[3] There was nothing else to be said; the word of a gentleman was not to be questioned.

In general, engaging in trade or a paid occupation would be an 'indiscretion' that negated any gentlemanly status that had been acquired by birth, because it imperilled freedom of thought and action. Nevertheless, there was an exception to the rule that a man could not be a gentleman if he was under an obligation, and that was if he engaged in one of the three medieval professions, or was serving his country in the army or navy. Only in this way could a man earn his living and yet remain a gentleman. However, during the eighteenth and early nineteenth centuries the function of the professions inverted, and from being the means by which a gentleman could follow a paid occupation whilst retaining his social status, they became the means by which those who were not gentlemen by birth could acquire that status by following a suitable occupation. Thus membership of a profession became a desideratum for social climbers.

In 1711 Joseph Addison still referred only to 'the three great Professions of Divinity, Law, and Physick',[4] and along with the armed services these remained the professions until the Royal Academy of Arts was founded in 1768 with the support of King George III. The purpose was to enhance the status and careers of artists, sculptors and architects; and royal sanction ensured that there was a corresponding enlargement of the activities that could be socially accepted as professions. This enlargement continued in the new century, when the industrial revolution brought into prominence the middle class, members of which could frequently earn an independent livelihood by acquiring and deploying the specialised knowledge that had become necessary in trade, industry and commerce. Naturally, many of them hankered after the social status offered by the existing professions, and so began to organise themselves along similar lines.

This was a social phenomenon of which the Victorians themselves were very much aware, and in 1888 Walter Besant, who had been eighteenth wrangler in 1859, wrote that during the preceding 50 years:

> There has been a great upward movement of the professional class. New professions have come into existence, and the old professions are more esteemed. It was formerly a poor and beggarly thing to belong to any other than the three learned professions; a barrister would not shake hands with a solicitor, a Nonconformist minister was not met in any society. Artists, writers, journalists, were considered Bohemians.[5]

Besant's comment suggests that early in the nineteenth century the social status enjoyed by Royal Academicians was not yet accorded to artists in general. The census of 1841 was the first to include a question concerning professions,[6] which by 1857 were of sufficient public interest to warrant a book on the subject; the author, H Byerley Thomson, wrote in the preface that he 'believes that this is the only general work on the English professions, and, as the first of its kind, with all diffidence, presents it for public consideration'.[7] He mentioned a number of new professions including, in the order that they were recognised, those of the artist, sculptor, architect, civil engineer and actuary; but he made no claim that mathematicians constituted a profession, and in this he has been followed for well over a century by subsequent writers on the subject.[8] So it is pertinent to enquire whether this is a justified omission by asking: what *were* the marks of a profession, and how was a profession distinguished from a mere trade, vocation or occupation? Could pure mathematics have constituted a profession?

Most of those who have made a study of the professions considered professionalisation in social terms. Although they never agree completely on their definitions, which are usually introduced ad hoc on each occasion, the differences tend to be in the detail, and one can distil from the literature a characterisation of the professions that has the additional advantage of reflecting their nature when they were developing in the second half of the nineteenth century.[9] The original connection of the professions with the medieval universities was forgotten, and the aspiring professionals tried to obtain their status by reproducing, as far as circumstances allowed, the social relations that existed between practitioners of the historic professions and the public.

From these historic professions were abstracted four principles that the emerging professions tried to adapt to their own requirements. First, ethics. A profession had codes of conduct that in any case of conflict overrode the particular private interests of the professionals. Thus they were deemed to be working under some higher moral law, in addition to the law of the land, and the public was assured that monetary considerations and self-interest were not the professionals' primary concerns, at least in theory. In this sense they sought to distinguish their motivations from those of industrialists, entrepreneurs and tradespeople, who in their defining forms were driven solely by the prospect of profit. The second principle was that of public service, in the sense that not only was the professional satisfying a demand, but also that it was in the public interest that the demand should be satisfied. The third characteristic was economic. The professionals sought to maintain standards by imposing entry requirements, and thereby implant in the public mind the idea of there being an elite group of individuals who alone were competent practitioners. Public-spirited though this may seem, it did of course have an additional effect that was of great benefit to the professionals, for by restricting entry in that way they were creating an artificial shortage of the particular expertise that they had to offer; the law of supply and demand therefore raised the economic rewards that were available. Thus the professionals were able to turn their particular knowledge or skill into capital, which yielded an income over and above the value of their services regarded purely in terms of the supply of labour.[10] Finally, in some cases the professions enjoyed a form of official recognition that gave them certain rights and privileges. For example, after the Pharmaceutical Society of Great Britain was incorporated in 1843 it was required by the Pharmacy Act of 1852 to maintain a register of suitably qualified pharmacists, for the purpose of preventing 'ignorant and incompetent persons from assuming the title of or pretending to be pharmaceutical chemists or pharmaceutists';[11] and legal recognition of the status of registered pharmacists was enshrined in the Pharmacy Act of 1868, under which they alone were allowed to sell poisons.

Implicit in these four characteristics is the need for some sort of formal specialised organisation, membership of which would determine who was, and who was not, a professional. Such an organisation could set and enforce entry standards, agree a code of ethics and professional behaviour, take punitive action against members who transgressed, and protect its members' interests. As regards the economic function, a profession was 'an occupation which so effectively controlled its labour market that it never had to behave like a trade union';[12] the distinction was one only of class and status.

Therefore one of the marks of social professionalisation was the existence of a professional body that could carry out all these functions, and if we are considering pure mathematics in Victorian Britain then the only candidate for such a body is the London Mathematical Society. From all that has been said in Chapter 3 concerning the Society's first half-century, it is possible to see how it measured up against the requirements. First, there was no formal entrance qualification: for the Council to approve membership it was only necessary for the applicant to find a proposer and two seconders from amongst the existing members, and pay the subscription; there was no requirement to have a degree, or indeed ever to have studied mathematics at all. Thus the level of mathematical attainment displayed by most of the members was accidental – by and large only able and enthusiastic mathematicians would have been interested in joining such a society. Of course, the requirement to have a proposer and two seconders was supposed to prevent the wrong sort of person from becoming a member, but there was no attempt to impose a standard of entry. Nor were members subject to any kind of code of ethics or behaviour, and there was no disciplinary procedure, apart from depriving a member of his membership if he failed to pay the subscription for several years. The Society did not lobby on behalf of its members, or encourage their employment; indeed, it made no claims for its members at all. In short, the Society bore none of the marks of a professional body.

Now we can turn to the individual mathematicians themselves, and ask whether their practice showed any marks of professional status. Of course, mathematics was used in many occupations, some of which were or became professions in their own right: for example, surveyors, actuaries and engineers all needed mathematical skill, and yet mathematics was not their profession; rather, they were professional surveyors, actuaries and engineers. Likewise, teachers of mathematics would require some mathematical expertise, and during the nineteenth century they gradually acquired professional status: Walter Besant once again contrasted the situation of teachers in the 1880s with that which had obtained 50 years before:

> There were a vast number of private schools. It was, indeed, recognised that when a man could do nothing else and had failed in everything that he had tried, a private school was still possible for him. ... The teaching of anything was held in contempt; to become a teacher was a confession of the direst poverty – there were thousands of poor girls eating out their hearts because they had to 'go out' as governesses.[13]

But as the century progressed and the lot of the teacher improved, so teachers of mathematics, whether at school or university, became professional teachers rather than professional pure mathematicians: inasmuch as they worked for the public good, it was not by researching pure mathematics, and inasmuch as pure mathematics was researched, it was not offered as a public good. All this makes it clear that pure mathematicians could not have constituted a social profession. This is not to say that pure mathematical research was never combined with other occupations, but it is to say that if a pure mathematician became a professional then it was in virtue of those other occupations rather than the practice of pure mathematics.

Far from presenting their members as valuable members of society, the LMS was quite ready to acknowledge that pure mathematicians made little contribution to the activities that their contemporaries held most dear. In 1880 the Society's president, Charles Watkins Merrifield, received a request from Mr H T Wood, the secretary of the Society for the Encouragement of Arts, Manufactures and Commerce, asking for suggestions as to who should receive the annual Albert Medal, first struck in 1864 to recognise 'distinguished merit in promoting Arts, Manufactures, or Commerce'. Merrifield, in reply, said that he had discussed the letter with Council,

> ... and they feel, with me, the high compliment paid to the Society by addressing this question to me as their President. With the exception of some of the distinguished men named in the list of those who have already received the medal, I cannot call to mind any name of a living person, of special repute as a mathematician, who has rendered such services to Arts Manufactures or Commerce, as to give him a title to the medal.[14]

Merrifield was restricted neither to his members nor to British citizens, because the award was made without regard to nationality; the second recipient had been the Emperor Napoleon III of France. Nevertheless, there was no one 'of special repute as a mathematician' whose name was worth putting forward. Furthermore, for Merrifield to refer to 'some of the distinguished men ... who have already received the medal' was marginally true at best, for there were only two to whom he could have been referring: George Airy, the Astronomer Royal, who was never a member of the LMS and received the medal in 1876, and William Thomson (later Lord Kelvin), who was a member and received the medal in 1879.[15]

The second variety of professionalisation was a process in which universities and similar institutions, rather than individuals, became the guarantors of knowledge. Until the Victorian age, authority derived more from the appeal and persuasiveness of authors' writings than from their credentials, and so was not necessarily specific to one particular field of expertise: polymaths could range freely over different disciplines without in any way compromising their relationship with the reader. Yet during the middle decades of the nineteenth century, when in many branches of knowledge there arose more complex and narrow specialities, authority began to be transferred from individuals to academic institutions that were defining disciplines more closely by creating departmental boundaries. Consequently, it was natural that the standing of authors often came to depend on academic credentials, and that they could no longer be so eclectic in their choice of subject matter: to use Ian Small's example, 'art critics and literary critics, such as John

Ruskin, could write freely on economics or politics in the 1860s, but could *not* do so in the 1880s'.[16]

It is this process – the shift in the site of intellectual authority from individuals to institutions – that offers a second definition of professionalisation. Whereas social professionalisation required the existence of an independent professional organisation, the alternative of academic professionalisation did not; rather, it applied to individuals who, in a relevant academic or institutional setting, were undertaking research in disciplines – such as history, philosophy and theology – that had no direct utility but were generally perceived as being of value in a civilised society. In this way, disciplines that did not qualify as social professions could nevertheless allow practitioners to enjoy the same status. However, this process did not extend to British pure mathematicians because, in contradistinction to most other forms of intellectual engagement, pure mathematics could not be presented in a form that laypeople could grasp, as illustrated by a newspaper report of a British Association meeting that ran: 'Saturday morning was devoted to pure mathematics, and so there was nothing of any general interest.'[17] Consequently, the public did not look to pure mathematicians for information or enlightenment, did not regard their work as being a contribution to civilised society, and was indifferent to their pronouncements. The Victorian age saw a marked decline in the proportion of workers in pure mathematics who were not in academic occupations – clergymen, lawyers, military men and so forth – but this did not lead to academic professionalisation, because its impact on the public's non-interest in pure mathematicians was negligible.

The British vision of pure mathematicians as being essentially isolated workers who had no need of any institutional or collegial support also militated against any tendency towards professionalisation; as Hardy was later to say, 'No mathematician in the world needs very much more than brains and leisure.'[18] This was particularly so because intuition was sometimes an element in a mathematical demonstration, and for that to be acceptable it had to be accompanied by confidence in the individual mathematician. Such a circumstance could arise when, faced with intractable algebraical or analytical complexity, a mathematician might formally prove a result only for particular instances, but state and use it more generally, a decision that required very fine judgement. Cayley had been adept at this and his intuition rarely let him down, although in one notable lapse he assumed for 12 years that all the members of a particular class of syzygies were independent, until Paul Gordan proved the contrary in 1868 using more general methods.[19]

The use of intuition as a ground for ascribing certainty to a result took several decades to die out, as we can understand from the disaster that befell Forsyth towards the end of his career. In 1934 he offered his newly completed two-volume *Intrinsic Geometry of Ideal Space* to his regular publisher Macmillan; his reputation ensured that the book was accepted by return of post, and the manuscript was not read or refereed before publication in 1935. The next year Professor Enrico Bompiani reviewed it very critically in *Nature*,[20] and elaborated his criticisms in a private communication to Forsyth. This led to an emotional letter from Forsyth to his publisher, written with 'unqualified regret', in which he admitted that he had not proved generally the fundamental theorem on which his book depended, but only in limiting instances from which he had nevertheless derived intuitive certainty as to its universal truth. He said:

> In each such [limiting] instance, the theorem is valid, beyond doubt: now I know that, beyond such instances, the theorem is not valid. Throughout the constructive analysis, the effect of one vitiating tacit assumption has persisted; it is of a kind that had never occurred in all my past experience. The error invalidates the fundamental theorem for all but the limiting instances, as well as all the inferences which are based on the general theorem when it is applied to non-limiting instances.[21]

Not only had he failed to prove the theorem for non-limiting instances, but also in all such instances it was in fact false, and this destroyed the value of much of the entire two volumes. Furthermore, his reference to 'all my past experience' suggests that he had relied on his intuition many times before, presumably with more success.

Evidently as late as the 1930s a distinguished British pure mathematician was prepared to go beyond the confines of a strictly logical development when it came to asserting mathematical truth, and this after more than two decades during which Hardy and Littlewood had been preaching as gospel the need for complete rigour at every stage of a mathematical demonstration. The problem was not with intuition as such, for to a considerable extent it directed a mathematician's research. Nor was it uncommon for an unproven result, expressed as an hypothesis, to be used to generate further results that would become proven only if the original hypothesis was itself proven; moreover, opinions differed as to what constituted a proof, which for Whitehead and Russell was merely such an abstract 'as is sufficient to convince a properly instructed mind'.[22] However, as Forsyth demonstrated, it was still possible for a mathematician to rely on an intuitive belief as the vital element of a theory, but without making this explicit, or perhaps even being consciously aware that there was an assumption that needed to be proved.

The evidence therefore argues against any assertion that pure mathematics, or pure mathematicians, 'professionalised' in either of its senses. As regards social professionalisation, there was no organisation that met the requirements, and mathematical practice had none of the marks of a profession. As regards academic professionalisation, pure mathematicians did not need their public authority to be endorsed by institutions because they had no public authority; and furthermore, pure mathematics was seen as an enterprise that was very personal to the mathematician. Therefore participation in the widespread process of professionalisation cannot be used as an off-the-shelf explanation of the discipline's ability to carve out a social and cultural niche for itself; something else was required. What that 'something else' might be exercised the minds of pure mathematicians from the 1840s onwards, and we can now consider the principal arguments that could be invoked to negate the criticisms that were directed at them and their discipline.

Earlier chapters have described how, during the first few decades of the nineteenth century, natural philosophy lost its unity when attention began to be focussed instead on the independence of the various sciences. At the same time, abstract Continental thought challenged the traditional view of mathematics as a fundamentally empirical discipline. Consequently the practice of pure mathematics in Britain became dissociated from mixed mathematics, because those practitioners who succumbed to Continental influences were no longer motivated to investigate nature and solve practical problems, nor did they consider natural processes to be the touchstones of mathematical truth. Such an approach was attractive to many mathematicians, and from early in the nineteenth century there was concern that enthusiasm for the new style of abstract, symbolic mathematics being introduced from the Continent would have a detrimental effect on scientific activity. In 1814 the architect

Charles Kelsall published plans for an ideal university, and was uncompromising in his opinions:

> One word on the mathematics. There is no study which, if cultivated with moderation, and solely in reference to *things here on earth*, has so beneficial an effect on the power of the mind; but none more pernicious, if deeply pursued through the mazes of algebra and analytics. Observation confirms this – All other sciences offer subjects of controversy, and different points of view, according as our genius and pursuits differ. Not so Mathesis. She sits in profound silence, and issues her incontrovertible edicts equally to the man of genius, and to the dunce. From her court there is no appeal. From which I deduce, as a corollary, that this science lays open no field for the communication of ideas. The higher walks of it, indeed, must be considered as an unsocial pursuit. And no harm, if they were left henceforth for those, who are better adapted for the order of La Trappe, than for mixing in society, and turning their knowledge to useful account.[23]

Kelsall had travelled widely on the Continent and was issuing a warning based on his experiences, rather than suggesting that getting lost in 'the mazes of algebra and analytics' had already held back the progress of humankind. However, during the following half-century the situation that he feared – a greatly increased output of pure mathematics – did arise, and many observers agreed with him that this was not an unmitigated good, as there seemed to be no justification for all the new activity. By 1845 we find Henry John Rose, an editor of the *Encyclopaedia Metropolitana*, writing of pure mathematics in his 'Introduction to the Pure Sciences':

> It is indeed almost an acknowledged fact, that, in some respects, we have a superfluity of knowledge of those principles. Our application of Mathematics to Natural Philosophy is so far from having exhausted all our stores of Pure Mathematics, that although there are still many problems too intricate for solution with our present means, yet there is also a large mass of results in pure mathematics which as yet have no specific application, and may be considered as stores reserved for future use.[24]

This suggestion – that the value of pure mathematics lay in its provision of 'stores reserved for future use' in natural philosophy – was open to two objections. First, the consequence would be that, as more and more excess pure mathematics was produced, so the marginal value of each new accretion would steadily diminish. Second, and more importantly, it ignored the motivation of the new breed of pure mathematicians, who did not work to produce a storehouse of methods and results that one day might possibly be of value in the applied sciences; such a role would have put them in a subordinate position, for they would have been there only to support scientists more directly engaged with empirical questions.

Such a justification for the pursuit of pure mathematics, even had it been plausible in 1845, was doomed to be nothing more than a stopgap measure. Indeed, by the 1850s it had run its course, for in 1858 James D Forbes published in the *Encyclopaedia Britannica* a lengthy dissertation entitled 'A Review of the Progress of Mathematical and Physical Science'; he was professor of natural philosophy at the University of Edinburgh, was noted for his work on heat conduction and glaciers, and had made significant contributions to the statistical analysis of scientific data. He did not treat of pure mathematics in detail, admitting that he lacked the expertise to do so, but he did write:

> Mathematicians have, since the time of D'Alembert, been noted for being more ready themselves to publish than become acquainted with what others have done; and one consequence of this has been the formation of a mathematical literature, able, profound, and original, but cumbrous, fragmentary, and full of repetitions. ... The fact is that a large proportion of the mathematical writings of even eminent authors are in a few years forgotten, or only casually consulted on some matter of history.[25]

D'Alembert died in 1783. The nature and tone of these comments by Rose and Forbes suggest that the tendency in mathematics that Kelsall had warned against was now in evidence, which was of particular concern to the many who believed that the continuing success of Britain as an internationally competitive industrial and commercial power required natural talents to be put to good use, rather than wasted in abstract and useless speculation. By 1864 – exactly half a century after Kelsall's warning – we find William Thomson writing to Hermann von Helmholtz:

> Oh! that the CAYLEYS would devote what skill they have to such things [as elasticity] instead of to pieces of algebra which possibly interest four people in the world, certainly not more, and possibly also only the one person who works. It is really too bad that they don't take their part in the advancement of the world, and leave the labour of mathematical solutions for people who would spend their time so much more usefully in experimenting.[26]

Further proof of the extent to which opinion had polarised can be gleaned from a rather tetchy exchange of correspondence in 1867 between two former senior wranglers: George Airy, the Astronomer Royal, and Arthur Cayley, who had occupied the Sadleirian chair at Cambridge since 1863. Although their debate centred on the value of pure mathematics as a pedagogic tool, Airy had some comments about the discipline that echoed the Trappist allusion of Kelsall:

> Now as to the Modern Geometry. With your praises of this science ... I entirely agree. And if men, after leaving Cambridge, were designed to shut themselves up in a cavern, they could have nothing better for their subjective amusement. ... But the persons who devote themselves to these subjects do thereby separate themselves from the world. They make no step towards natural science or utilitarian science, the two subjects which the world specially desires. ... I believe that many of the best men of the nation consider that a great deal of time is lost on subjects which they esteem as puerile, and that much of that time might be employed on noble and useful science. ... So much time is swallowed up by the forced study of the Pure Mathematics that it is not easy to find anybody who can really enter on these subjects in which men of science want assistance.

It is not our purpose here to enquire about his evidence; what matters is that his views reflected a widespread opinion that the pure mathematician was an anchorite, one who had deliberately and selfishly cut himself off from the concerns of the rest of humanity in order to study something of interest and value only to himself or, at best, to very few others. Under the old tradition, pure mathematics had been whatever in mixed mathematics could be studied in isolation from the empirical aspects, but Airy had identified the increasingly abstruse, abstract and generalised nature that pure mathematics was acquiring. He gave an example:

> I do not know that one branch of Pure Mathematics can be considered higher than another, except in the utility of the power which it gives. Measured thus, the Partial Equations are very useful and therefore stand very high, as far as the Second Order. They apply, up to that point, in the most important way, to the great problems of nature concerning *time*, and *infinite division of matter*, and *space*: and are worthy of the most careful study. Beyond that order they apply to nothing.

Cayley offered a curiously oblique refutation that hardly answered the point, but whatever the merits of the opposing arguments, a clear distinction was drawn between mathematicians who offered practical assistance to 'men of science' by developing and applying pure-mathematical results and techniques – these mathematicians had value – and mathematicians who by inclination or training chose not to do so – these mathematicians had no value.[27]

Something over a year later Thomas Henry Huxley, a far more pungent controversialist than Airy, took up the criticism of pure mathematics. In 1869, when attacking Comtian positivism, he claimed that pure mathematics 'is that which knows nothing of observation, nothing of experiment, nothing of induction, nothing of causation',[28] and later said that 'mathematical training is almost purely deductive. The mathematician starts with a few simple propositions the proof of which is so obvious that they are called self-evident, and the rest of his work consists of subtle deductions from them.'[29] These comments aroused the ire of James Joseph Sylvester, who was just coming to the end of a 15-year term as professor of mathematics at the Royal Military Academy, Woolwich. He was not only one of the foremost British mathematicians of the day, but also a poet, or at least a versifier, and the author of a book entitled *The Laws of Verse*, to which he attached great importance. He had been the first practising Jew to matriculate at a Cambridge undergraduate college,[30] and although second wrangler in 1837 he could not take his degree because of his faith; he was eventually awarded a BA in 1872 when there were no longer any relevant religious tests. In his public speaking and lecturing he could be relied upon to use excitable and colourful language to get his points across, and he was therefore an ideal candidate for refuting Huxley's opinion of pure mathematics.[31] Sylvester decided to reply on a very public occasion, namely when addressing Section A of the British Association meeting to be held later that year, and it was entirely in character that he evoked vivid imagery that was to have an abiding influence on the characterisation of the pure mathematician.

Sylvester began by acknowledging that mathematical reasoning was perceived as being arid and sterile, and that mathematicians were believed to be devoid of imagination and creativity. That this had been a widespread perception for many years can be understood from Edgar Allan Poe's short story 'The Purloined Letter', first published in 1845, which presented such a perception in a popular form. Poe's starting point was the assumption, apparent in Huxley's comments, that mathematical reasoning, while providing the model for reasoning in general, was nevertheless nothing more than a mechanical process. With this in mind he allowed Dupin (a kind of French prototype of Sherlock Holmes) to solve a mystery by crediting the villain – a mathematician, poet and minister of state – with a creative imagination, which the prefect of police had assumed that no mathematician could possess; Dupin, on the other hand, believed it to be necessary for true reasoning. After a disquisition by Dupin on the folly of mathematicians in seeking to elevate mathematical truths (such as that the whole is equal to the sum of its parts) to general truths applicable to morals and conduct, the following conversation ensues – Dupin's confused friend, the narrator, speaks first:

> 'But is this really the poet?' I asked. 'There are two brothers, I know; and both have attained reputation in letters. The minister, I believe, has written learnedly on the Differential Calculus. He is a mathematician and no poet.'
>
> 'You are mistaken; I know him well; he is both. As poet *and* mathematician he would reason well; as mere mathematician he could not have reasoned at all, and thus would have been at the mercy of the Prefect.'
>
> 'You surprise me,' I said, 'by these opinions, which have been contradicted by the voice of the world. You do not mean to set at naught the well-digested idea of centuries. The mathematical reason has long been regarded as *the* reason *par excellence*.'[32]

We do not know whether Sylvester – like the minister, both mathematician and poet – had read 'The Purloined Letter', but he was keen to contradict the idea, espoused by Dupin's friend and proclaimed by Huxley, that the thought processes of pure mathematicians were entirely deductive, involved no imagination or creativity, and so were quite different from the reasoning of poets, men of affairs, and those who were increasingly being referred to as scientists. Sylvester summarised in typical fashion the argument that he was opposing by saying:

> [rather than] favour the notion that we Algebraists (who regard each other as the flower and salt of the earth) are a set of mere calculating machines endowed with organs of locomotion, or, at best, a sort of poor visionary dumb creatures only capable of communicating by signs and symbols with the outer world, I have resolved to take heart of grace and to say a few words, which, I hope to render, if not interesting, at least intelligible, on a subject to which a large part of my life has been devoted.

After quoting from Huxley's two papers, Sylvester went on to refute him completely:

> I think no statement could have been made more opposite to the undoubted facts of the case, that mathematical analysis is constantly invoking the aid of new principles, new ideas, and new methods, not capable of being defined by any form of words, but springing direct from the inherent powers and activity of the human mind, and from continually renewed introspection of that inner world of thought of which the phenomena are as varied and require as close attention to discern as those of the outer physical world ... that it is unceasingly calling forth the faculties of observation and comparison, that one of its principal weapons is induction, that it has frequent recourse to experimental trial and verification, and that it affords a boundless scope for the exercise of the highest efforts of imagination and invention.

Sylvester wanted to claim that pure mathematicians, although unmotivated by empirical considerations, were engaged in the same kind of mental activity as scientists, and he introduced as being applicable to pure mathematics some of the key words and phrases that Huxley and his followers applied to the sciences: 'observation', 'induction', 'experimental trial and verification' and 'invention'. However, Sylvester had also admitted that pure mathematics was carried on entirely in the mind, and he therefore needed to make more definite his characterisation of pure-mathematical activity. Scientists' activities were concerned with the observable world, but with what were pure mathematicians' activities concerned?

In attempting to answer this question Sylvester invoked mathematical Platonism, in which the mind of the mathematician apprehends an independently existing world of mathematical objects. To make this more persuasive, he couched his argument in terms of natural history, the discipline in which Huxley had made his reputation; indeed, only a few moments earlier – doubtless mindful of the combative Huxley's reputation as 'Darwin's bulldog', and perhaps seeing himself in a similar role – he had referred to 'our Cayley, the central luminary, the Darwin of the English school of mathematicians'. Sylvester took as an illustration the algebraic theory of invariants, of which Cayley and he were the leading British exponents. He asked:

> How did this originate? In the accidental observation by Eisenstein, some score or more years ago, of a single invariant ... which he met with in the course of certain researches just as accidentally and unexpectedly as M. Du Chaillu might meet a Gorilla in the country of the Fantees, or any one of us in London a White Polar Bear escaped from the Zoological Gardens. Fortunately he pounced down upon his prey and preserved it for the contemplation and study of future mathematicians. It occupies only part of a page in his collected posthumous works. This single result of observation (as well entitled to be so called as the discovery of Globigerinae in chalk or of the Confoco-ellipsoidal structure of the shells of the Foraminifera), which remained unproductive in the hands of its distinguished author, has served to set in motion a train of thought and to propagate an impulse which have led to a complete revolution in the whole aspect of modern analysis, and whose consequences will continue to be felt until Mathematics are forgotten and British Associations meet no more.[33]

Sylvester was here offering a succession of images that appealed directly to those aspects of scientific activity championed by Huxley, who had himself undertaken research into globigerinae and foraminifera, and was therefore clearly a target. The reference to a polar bear remains

Figure 6.1 First catch your invariant: Sylvester's characterisation of mathematical discovery.

a mystery, at least to the present author, but the reference to 'a Gorilla in the country of the Fantees' was a very specific attempt to ally pure mathematics with geographical exploration. Paul Du Chaillu was an explorer in Africa who a few years earlier had become a considerable public figure after publishing popular accounts of his travels. They included a dramatic story of how in Fan country (now part of Gabon) he became the first white man to come face to face with a gorilla; he even provided an engraving of the event (Figure 6.1).[34]

It was this image that Sylvester knew would be familiar to many in his audience – even those who were unable immediately to call to mind the confoco-ellipsoidal structure of the shells of the foraminifera – and that there would be thus impressed on them in a very vivid manner the idea of the pure mathematician as an explorer in a strange and uncharted world, a world full of things. To the pure mathematician, Sylvester was saying, the mathematical world and the things in it such as invariants are as real as Africa and the gorilla were to Du Chaillu, and a pure mathematician's business is to discover, report upon and capture for detailed examination all the things that this world contains.

Sylvester thereby gave great public exposure to the characterisation of the pure mathematician as an explorer, and to the view that much of the vocabulary used by Huxley to characterise science could also be applied to pure mathematics. This stance has remained a popular one with mathematicians, and in 1940 was given somewhat less colourful expression by Hardy when he wrote: 'I believe that mathematical reality lies outside us, that our function is to discover or *observe* it, and that the theorems that we prove, and which we describe grandiloquently as our "creations", are simply notes of our observations.'[35] However, for the Victorians this presentation of the pure mathematician in the guise of an explorer and natural historian was incomplete, although perhaps persuasive as far as it went. The justification for expending time, money and effort in exploring Africa was clear – it increased scientific knowledge, furthered the ambitions of Empire and brought salvation to the heathen, and was therefore a universal good – but what was the public benefit in exploring the world of pure mathematics?

An attempt to answer that question was made in 1876 by Henry Smith in his presidential address to the London Mathematical Society. In this he offered pure mathematics as a necessary concomitant of a vigorous and expanding Empire, saying that:

> I should not wish to use words which may seem to reach too far, but I often find the conviction forced upon me that the increase in mathematical knowledge is a necessary condition for the advancement of science, and if so, a no less necessary condition for the improvement of mankind. I could not auger well for the enduring intellectual strength of any nation of men, where education was not based on a solid foundation of mathematical learning, and whose scientific conceptions, or in other words whose notions of the world and of the things in it, were not braced and girt together with a strong framework of mathematical reasoning. It is something for men to learn what proof is and what it is not; and I do not know where this lesson can be better learned than in the schools of a science which has never had to take one footstep backward, which has never asserted without proof, or retracted a proved assertion; a science which, while ever advancing with human civilization, is as unchangeable in its principles as human reason; the same at all times and in all places; so that the work done at Alexandria or Syracuse two thousand years ago (whatever may have been added to it since) is so perfect in its kind, and as direct and unerring in its

appeal to our intelligence, as if it had been done yesterday at Berlin or Göttingen by one of our own contemporaries. Perhaps also it might not be impossible to show, and even from instances within our own time, that a decline in the mathematical productiveness of a people implies a decline in intellectual force along the whole line; and it might not be absurd to contend that on this ground the maintenance of a high standard of mathematical attainment among scientific men of a country is an object of almost national concern.[36]

Smith is here echoing, although for different reasons, the Cambridge Royal Commissioners who said in their 1852 Report that 'On the importance of [mathematical] studies it is impossible to dwell too strongly ... it is impossible not to feel that the glory of the country itself is intimately bound up with our national progress in this direction ...'.[a] He was, however, considerably more prolix, for reasons given in 1902 by Alexander Macfarlane when he said that Smith sometimes expressed his 'anti-utilitarian view of mathematical science' in 'exaggerated terms as a defiance to the grossly utilitarian views then popular'.[37] Certainly the serpentine sentence beginning 'It is something for men to learn ...' conveys the impression that mathematics has always been above controversy or doubt, but this is completely untrue, as Smith must have known perfectly well.

There are two other points to note about this passage. The first is that Smith was not concerned with mathematics as a pedagogical tool. He was addressing the London Mathematical Society, which by this time had experienced at the hands of James Wilson the unwelcome dissension that such discussions could bring. Pedagogy was, of course, continually being discussed at Cambridge, usually with a good deal of heat, but the Society now saw itself as a haven of tranquillity in which could be nurtured the vision of the pure mathematician as one unencumbered by uncertainties. Furthermore, Smith had spent his career at Oxford, where mathematics, like the classics, was treated far less competitively than at Cambridge. Thus his view of mathematics had not been coloured by its being a vehicle for intense pedagogical competition – a purpose that would have suggested itself naturally to a Cambridge man, for at Cambridge mathematical education

---

[a] See p. 35.

concentrated on the rapid solving of short questions that centred around the requirements of Newtonian natural philosophy. Such an approach was well-understood by the public, and expressed in Bernard Shaw's play *Mrs Warren's Profession*, which he wrote in 1893. In Act I, Vivie Warren, a young woman who has recently been placed equal to the third wrangler in the Tripos, explains:

> Do you know what the mathematical tripos means? It means grind, grind, grind for six to eight hours a day at mathematics, and nothing but mathematics. I'm supposed to know something about science; but I know nothing except the mathematics it involves. I can make calculations for engineers, electricians, insurance companies, and so on; but I know next to nothing about engineering or electricity or insurance. I don't even know arithmetic well. Outside mathematics, lawn-tennis, eating, sleeping, cycling, and walking, I'm a more ignorant barbarian than any woman could possibly be who hadn't gone in for the tripos.

This described, with only a small degree of dramatic licence, a competitive process that was completely outside Smith's experience as a student at Oxford.

The second point to note concerning Smith's address is that his use of the phrase 'and even from instances within our own time' – implying that such instances were not in the main line of his reasoning – suggests that his fundamental approach was based on an induction from historical examples. He offered no causative mechanism to explain why a healthy degree of activity in pure mathematics was necessary for a prosperous and advancing nation, but simply said that there was in fact a strong correlation between the two, and that we would be wise to learn the lesson of history. As a rhetorical flourish to bolster the self-confidence of the Society's members this was doubtless acceptable, but it was unlikely to satisfy the holders of 'the grossly utilitarian views then popular'.

To conclude this survey of mathematical apologias: we have seen that professionalisation cannot be adduced as an explanation for pure mathematicians' ultimate success in establishing their discipline. We have also considered the three principal arguments by which pure mathematicians sought ineffectually to rebut the criticism that their activities constituted a selfish refusal to provide what the industrial revolution most needed, namely technical expertise. By the 1850s the argument that they were providing a store of methods and results for future use in the applied sciences had already run out of steam, in the face of the sheer quantity of pure mathematics that was being produced, and the implicit denial of pure mathematicians' true motives. The ingenious attempt to portray the pure mathematician in the same light as the natural historian, exploring and reporting on the immaterial but real world of mathematics, can still be encountered today, but it left unanswered vital questions concerning the value of this activity for society in general. Moreover, the suggestion that vigorous pure mathematical activity was necessary for the advancement of humankind was not an argument at all, but merely an assertion backed up by misstated facts. Accordingly, by the late 1870s it was clear that for pure mathematicians to have any credibility in the eyes of the world at large, they needed to present their discipline in a completely new way.

Notes and References

1. See, for example, I Grattan-Guinness, 'Introduction', in I Grattan-Guinness (ed.), *Companion Encyclopedia of the History and Philosophy of the Mathematical Sciences*, 2 (London: Routledge, 1994), p. 171; Adrian Rice and Robin J Wilson, 'The Rise of British Analysis in the Early 20th Century: The Role of G H Hardy and the London Mathematical Society', *Historia Mathematica*, 30, 2 (2003), 173–194, at 174.
2. Robert Boyle, 'The Gentleman': manuscript in the Boyle Papers of the Royal Society, 37, fos. 160–163, quoted in Steven Shapin, *A Social History of Truth: Civility and Science in Seventeenth-Century England* (Chicago: University of Chicago Press, 1994), p. 146.
3. Letters from Leibniz to Oldenburg dated 8 March 1673, and from Oldenburg to Leibniz dated 16 May 1673, in A R Hall and M B Hall (eds. & trs.), *The Correspondence of Henry Oldenburg*, IX (Madison, WI: University of Wisconsin Press, 1973), pp. 494, 668.
4. *The Spectator*, 24 March 1711.
5. Walter Besant, *Fifty Years Ago* (London: Chatto & Windus, 1888), p. 262.
6. Ian Small, *Conditions for Criticism: Authority, Knowledge, and Literature in the Late Nineteenth Century* (Oxford: Clarendon Press, 1991), p. 20.
7. H Byerley Thomson, *The Choice of a Profession. A Concise Account and Comparative Review of the English Professions* (London: Chapman and Hall, 1857), p. vi.
8. For example: Francis Davenant, *What shall My Son be? Hints to Parents on the Choice of a Profession or Trade; and Counsels to Young Men on their Entrance into Active Life. Illustrated by Anecdotes and Maxims of Distinguished Men. With a Copious Appendix of Examination Papers and other Practical Information* (London: Partridge, 1870); A M Carr-Saunders and P A Wilson, *The Professions* (Oxford: Clarendon Press, 1933); Geoffrey Millerson, *The Qualifying Associations:*

A *Study in Professionalization* (London: Routledge, 1964); W J Reader, *Professional Men: The Rise of the Professional Classes in Nineteenth-Century England* (London: Weidenfeld & Nicolson, 1966); Magali Sarfatti Larson, *The Rise of Professionalism: A Sociological Analysis* (Berkeley, CA: University of California Press, 1977); Harold Perkin, *The Rise of Professional Society: England since 1880* (London: Routledge, 1989).

9. Morris L Cogan, 'The Problem of Defining a Profession', *Annals of the American Academy of Political and Social Science*, 297 (1955), 105–111.
10. See the discussion of this in Perkin, *The Rise of Professional Society*, pp. 7–8.
11. The Pharmacy Act, 1852, Preamble.
12. Perkin, *The Rise of Professional Society*, p. 23.
13. Besant, *Fifty Years Ago*, pp. 154, 263.
14. Letters from Wood to Merrifield dated 6 April 1880 and Merrifield to Wood dated 10 April 1880; both loose in the Council Minute Book at those dates. The minutes of the Council meeting on 8 April 1880 do not record the discussion.
15. For a list of recipients and their achievements, see Henry Trueman Wood (ed.), *Directory of the Royal Society of Arts* (London: Bell, 1909), pp. 43–47.
16. Small, *Conditions for Criticism*, p. 43.
17. A R Forsyth, 'Presidential Address to Section A of the British Association', *Report of the Sixty-seventh Meeting of the British Association for the Advancement of Science* [1897] (London: John Murray, 1898), pp. 541–549, at 542.
18. G H Hardy, 'Mathematics', *The Oxford Magazine*, 48 (5 June 1930), 820.
19. Karen Hunger Parshall and David E Rowe, *The Emergence of the American Mathematical Research Community, 1876–1900: J J Sylvester, Felix Klein, and E H Moore* (Providence, RI: American Mathematical Society and The London Mathematical Society, 1994), p. 104.
20. Enrico Bompiani, 'Intrinsic Geometry of Ideal Space', *Nature*, 138, 3487 (1936), 343–344.
21. Letters between Forsyth and Macmillan & Co dated 6 and 8 March 1934 and 24 September 1936: Macmillan Archives (British Library), Add 55197, fos. 133–136.
22. A N Whitehead and Bertrand Russell, *Principia Mathematica*, 1 (Cambridge: Cambridge University Press, 1910), p. 3.
23. Charles Kelsall, *Phantasm of an University: With Prolegomena* (London: White, Cochrane, 1814), p. 29.
24. Henry John Rose, 'Introduction to the Pure Sciences', in *Encyclopaedia Metropolitana, or, Universal Dictionary of Knowledge, on an Original Plan: comprising the Twofold Advantage of a Philosophical and an Alphabetical Arrangement with appropriate Engravings*, I (Pure Sciences I) (London: B. Fellowes, 1845), p. xii. The introduction is signed 'HJR', but comments on p. xxi show that it was written by Henry John Rose and not his elder brother Hugh James Rose, who was also an editor of the *Encyclopaedia Metropolitana*.
25. James D Forbes, *A Review of the Progress of Mathematical and Physical Science in more Recent Times, and particularly between the Years 1775 and 1850; being one of the Dissertations prefixed to the Eighth Edition of the Encyclopaedia Britannica* (Edinburgh: A & C Black, 1858), p. 6.
26. Letter from Thomson to Helmholtz dated 31 July 1864, in Silvanus P Thompson, *The Life of William Thomson Baron Kelvin of Largs*, I (London: Macmillan, 1910), p. 433; Thomson's capitalisation.
27. Letters from Airy to Cayley dated 9 December 1867 and 8 November 1867, and from Cayley to Airy dated 6 December 1867, in Wilfrid Airy (ed.), *Autobiography of*

213 / Apologias for Pure Mathematicians

Sir George Biddle Airy, K.C.B., M.A., LL.D., D.C.L., F.R.S., F.R.A.S., Honorary Fellow of Trinity College, Cambridge, Astronomer Royal from 1836 to 1881 (Cambridge: Cambridge University Press, 1896), pp. 273–278.

28. T H Huxley, 'The Scientific Aspects of Positivism', *The Fortnightly Review*, 5, 30 (1869) 653–670, at 667.
29. T H Huxley, 'Scientific Education: Notes of an After-Dinner Speech', *Macmillan's Magazine*, 20, 116 (1869), 177–184, at 182.
30. William Frankel and Harvey Miller (eds.), *Gown and Tallith: In Commemoration of the Fiftieth Anniversary of the Founding of The Cambridge University Jewish Society* (London: Harvey Miller, 1989), p. 202.
31. The standard biography of Sylvester is Karen Hunger Parshall, *James Joseph Sylvester: Jewish Mathematician in a Victorian World* (Baltimore MD: Johns Hopkins, 2006).
32. Edgar Allan Poe, 'The Purloined Letter', in William P Trent (ed.), *The Gold Bug, The Purloined Letter and Other Tales* (Boston MA: Houghton Mifflin, 1898), p. 66. First published in *The Gift* for 1845.
33. Quotations from J J Sylvester's 'Address', in *Report of the Thirty-ninth Meeting of the British Association for the Advancement of Science* [1869] (London: John Murray, 1870), pp. 1–9 of 'Notices and Abstracts'.
34. Paul B Du Chaillu, *Explorations & Adventures in Equatorial Africa; with Accounts of the Manners and Customs of the People, and of the Chace of the Gorilla, Crocodile, Leopard, Elephant, Hippopotamus, and other Animals* (London: John Murray, 1861), between pp. 70 and 71.
35. G H Hardy, *A Mathematician's Apology* (Cambridge: Cambridge University Press, 1967), pp. 123–124; originally published 1940.
36. H J Stephen Smith, 'On the Present State and Prospects of some Branches of Pure Mathematics', *Proceedings of the London Mathematical Society*, VIII (1876), 6–29, at 28–29.
37. Alexander Macfarlane, *Lectures on Ten British Mathematicians* (New York: Wiley, 1916), p. 100; the lecture was delivered in 1902.

7 EMBRACING BEAUTY

So far, all the contemporary commentators have presented Victorian pure mathematics as having the nature of a science, for although they may have disagreed as to its value, they did agree that its purpose was to produce knowledge of some kind. What was at issue was the nature of that knowledge; whether, and if so why, it was worth possessing; and whether pure mathematicians would have been better employed in some more useful enterprise. However, it is difficult to see how the debate could have continued along these lines, and the argument presented here is that it did not; that instead there appeared a new characterisation of the pure mathematician as an artist, a creator of something beautiful; and that this characterisation was given its cogency by contemporary developments in the world of art. But first it is necessary to explain the background that makes such an argument possible.

Beginning in the late 1860s, the Oxford critic Walter Pater wrote several essays on aesthetics that were collected and published in 1873 under the title *Studies in the History of the Renaissance*; the book was popular for such a work, and by 1890 it had sold more than 4,000 copies.[1] In these essays he called into question some of the most cherished Victorian beliefs by denying that the value of a work of art lay in the moral truth that it conveyed. Conventional opinion held that such truth was manifested in the work of art as beauty, the contemplation of which led to the moral improvement of the individual; consequently the role of aesthetics was to define the nature of beauty and the manner in which moral truth was conveyed. However, Pater regarded all such speculation as pointless, arguing instead that a thing was beautiful only in virtue of inducing an intense feeling of pleasure, and that beauty could be found anywhere, not just in traditional art and literature but also in 'the fairer forms of nature and human life'. He turned the focus of attention away from metaphysics and towards the experiences of the beholder in the presence of a specific beautiful thing: 'To define beauty not in the most abstract, but in the most concrete terms possible, not to find a universal formula for it, but the formula which expresses most adequately this or that special manifestation of it, is the aim of the true student of aesthetics.' The questions to ask were: 'What is this song or picture, this engaging personality presented in life or in a book, to *me*? What effect does it really produce on me? Does it give me pleasure? and if so, what sort or degree of pleasure? How is my nature modified by its presence and under its influence?' The object of life, he said, was to experience as much of this aesthetic pleasure as possible, and he claimed that 'art comes to you professing frankly to give nothing but the highest quality to your moments as they pass, and simply for those moments' sake'.[2] This final quotation is taken from the highly controversial Conclusion to the *Studies*, which he omitted from the second edition because its sentiments were considered too dangerous for the young men who might read them.[3] Pater's view of aesthetics was summed up in the phrase 'art for art's sake', usually attributed to the French philosopher Victor Cousin, but which Pater made familiar in Britain.

Pater's concentration on the experiences of the individual when contemplating a specific beautiful thing widened the scope of aesthetics beyond the traditional fine arts and literature, for its new mission was to examine, not beauty in the abstract, but the personal response that a particular manifestation of beauty engendered. Pater had originally considered a career in the church, but his argument offended religious opinion because he considered that a life spent in pursuit of beauty – that is, a life spent in garnering as much aesthetic pleasure as possible – was a life well spent. Although a quiet Oxford don who lived a life of unassuming rectitude only marginally touched by scandal, he had lost all sense of religious vocation by the time he came to develop his theory of aesthetics.

Soon Pater's doctrines were enthusiastically adopted by a flamboyant group of artists and writers led by the painter James McNeill Whistler and, later, by the poet, essayist and playwright Oscar Wilde. This group became known as 'aesthetes'; and whereas Pater's aesthetics was wholly intellectual, the aesthetes tried to express their doctrines in their everyday lives. They believed that some people – namely, themselves – had superior powers of perception, and it was only those possessing such powers who could spend their lives in the contemplation of beauty.[4] In Ian Small's words: 'For the serious apologist for Aestheticism, the experience of beauty did not simply *contrast* with the prosaic experiences or realities of ordinary life: it was an experience of a quite different order. Experience of the "real" world ... was, as it were, suspended or superseded by the more vivid, and perceptually distinct, "world" of art.'[5] For those who did possess such powers of perception, beauty could be found not just in traditional works of art but in apparently mundane objects, some of which, such as lilies, peacocks' feathers, sunflowers and blue china, became iconic.[6] The aesthetes tried to organise their lives, their conversation, their dress and their friends so as to maximise their exposure to beauty and hence to pleasure; and this naturally meant shunning the unpleasant and vulgar. Indeed, Wilde's flippancy on this subject is often considered to have caused the tide of opinion to turn firmly against him in the action for libel that he brought against the Marquess of Queensberry in 1895. Wilde was asked, in what H Montgomery Hyde refers to as 'the climax to the cross-examination', whether he had ever kissed Walter Grainger, a servant. He replied: 'Oh, dear no. He was a peculiarly plain boy. He was, unfortunately, extremely ugly. I pitied him for it.'[7] This was not, of course, the usual reason that Victorian gentlemen refrained from kissing their male servants.

The word 'aestheticism' first arose in the middle of the nineteenth century, and as R V Johnson described in his historical study, it 'denoted something new: not merely a devotion to beauty, but a new conviction of the importance of beauty as compared with – and even in opposition to – other values'.[8] These 'other values' included those that were responsible for the general move in society towards professionalisation, so it was not only the pure mathematicians who failed to participate in this trend; the aesthetes did too:

> Much of the point of professionalization and specialization in the last three decades of the nineteenth century exists precisely in the marking out of specific and exclusive objects of study for each discipline. The deliberate 'failure', or unwillingness, of Wilde or Pater to talk about the object of study of their particular discipline – the art-object – was also a point-blank refusal to be party to the newly specialized academic world.[9]

The aesthetic tendency had many critics, for to deny the moral function of art, and to assert instead the value of a life spent in the pursuit of beauty and pleasure, was deeply offensive to many Victorians.[a] The most prominent of these critics was probably John Ruskin, who had a very elevated view of art, and believed that its purpose was to enhance morality and religious feeling. His opposition to the new doctrine of aestheticism attracted much publicity after he intemperately wrote of one of Whistler's paintings that he never expected to find an artist demanding 200 guineas 'for throwing a pot of paint in the public's face'. Whistler successfully sued for libel, but the jury's opinion of his art was shown by the damages that it awarded, which amounted to one farthing (approximately one-tenth of a penny in today's currency).

However, these debates were not carried on only in higher intellectual and academic circles; they touched all classes of society. The aesthetes' adopted lifestyle made them prime targets for lampooning, and Gilbert and Sullivan's comic opera *Patience, or Bunthorne's Bride*, which began a run of over 400 performances in April 1881,[10] was aimed four-square at them, with the central character of Reginald Bunthorne being modelled on Whistler. Although black-haired, Whistler had some locks of white, and in the opening production George Grossmith, playing Bunthorne, sported a streak of white hair that made it obvious who was being mocked. Henry Lytton, who was later to play Bunthorne many times, wrote:

> When Mrs. D'Oyly Carte [the wife of the producer] first saw me as *Bunthorne*, she exclaimed 'How you do remind me of Whistler!' That was a compliment; it was from Whistler, of course, that this role was understood to be drawn, and so I was not loath to copy the poet's photograph, even to the white locks in his ample jet-black hair![11]

It is sometimes thought that Bunthorne ('a fleshly poet', according to the stage directions) was modelled on Wilde, but this is incorrect, and indeed when *Patience* opened Wilde did not have the notoriety that he was later to acquire; it was Whistler who was the recognisable public figure. Note that Lytton, in referring to 'the poet's photograph', is confusing Whistler with his dramatic alter ego.

---

[a] Although aestheticism is usually spoken of as a 'movement', the present author follows Johnson in characterising it 'as a current of taste and ideas, as a "tendency" rather than a "movement."' (Johnson, *Aestheticism*, p. 46).

Gilbert's lyrics for *Patience* played on the aesthetes' belief that their intellect operated on a plane well above that of the man in the street; so when Bunthorne, unobserved, favours the audience with a candid appraisal of an aesthete's state of mind, he sings:

> And ev'ry one will say
> As you walk your mystic way,
> 'If this young man expresses himself in terms too deep for me,
> Why, what a very singularly deep young man this deep young man must be!'

He optimistically expected that the elevated nature of the aesthetes' discourse would increase the awe with which they were regarded by an uncomprehending public.

It was not only on stage that the aesthetes were satirised, because in the popular weekly magazine *Punch* they were lampooned in cartoons such as that by George du Maurier shown in Figure 7.1, which was published in 1880. The central character, Jellaby Postlethwaite, is using the faculty of perception peculiar to the aesthetes to demonstrate very publicly the difference between himself and the common people who surround him. *They* need an intake of food at the luncheon hour, but *he* needs only to feed his soul by the contemplation of a lily, an activity that ordinary mortals such as the waiter, the onlookers, and indeed du Maurier himself, find absurd.

The lampoons by du Maurier and the satire of W S Gilbert were directed at the personal excesses of some of the aesthetes, but they also conveyed the aesthetic creed to many who were not directly familiar with the work of Walter Pater and his followers. The wider public thereby became aware that the moralistic view of art and artists was not the only one; there was an intellectually respectable alternative, which centred on four main points. First, there is no restriction on what may be counted as beautiful; second, some people have a greater ability than others to perceive beauty; third, a thing is beautiful in virtue of the pleasure that it gives to those who contemplate it; and fourth, the object of life should be to obtain as much of this pleasure as possible.

The developments in contemporary thought that were inspired by Pater's work on aesthetics took place during the 1870s, and shortly afterwards there appeared the first signs that pure mathematicians were ready to adapt these new ideas for their own purposes. In 1882 the Council of the British Association nominated Arthur Cayley, the

AN ÆSTHETIC MIDDAY MEAL.

At the Luncheon hour, Jellaby Postlethwaite enters a Pastrycook's and calls for a glass of Water, into which he puts a freshly-cut Lily, and loses himself in contemplation thereof.

Waiter. "SHALL I BRING YOU ANYTHING ELSE, SIR?"
Jellaby Postlethwaite. "THANKS, NO! I HAVE ALL I REQUIRE, AND SHALL SOON HAVE DONE!"

Figure 7.1 Cartoon from *Punch* by George du Maurier, published in 1880, lampooning the aesthetes.

Sadleirian Professor at Cambridge, as president for the following year; but for the first time the presidency was to be contested. The venue for the 1883 meeting had originally been Oxford, but some Oxford academics thought that Cayley's nomination unfairly failed to recognise the extent to which their university had embraced the sciences since the Royal Commission of 1852, and so they put forward Lord Salisbury, the Chancellor, as an alternative candidate. Cayley's nomination was eventually confirmed, but there remained some bitterness and the meeting was moved to Southport.[12] As a result, not only was Cayley to be president, and so a public figure – a rare event for a pure mathematician – but also his controversial elevation had generated even more interest in the meeting than was usual.

It was therefore unsurprising that in September 1883, shortly before the meeting, the noted Irish geometer and divine George Salmon felt the need to publish in the popular science journal *Nature* an appreciation of his friend Cayley, who had been so suddenly thrust into the public eye. *Nature* had a large circulation, so although Cayley was Britain's foremost algebraist, Salmon could not write as if for a readership of mathematicians who required only to have their own views confirmed; rather, he had to convey to non-mathematicians just why he believed that Cayley was an important figure. Some of Cayley's work did have applications, but because he was generally considered to be the country's foremost pure mathematician Salmon could not allow himself to pursue the 'servant of the sciences' argument. He did spend time trying to explain some of Cayley's mathematics, but had to face the uncomfortable truth that this would be incomprehensible to most of his readers.

Salmon was well aware of the difficulties, for he wrote at the outset that 'the quiet life of a student' – he meant Cayley – 'is not likely to be rich in sensational incidents, and of the nature of the work done by a labourer in the field of pure mathematics it is not possible to give more than a vague idea to the outside world'.[13] In this passage Salmon created a clear divide between scientists and pure mathematicians, for it is difficult to think of any science of which a Victorian would have said that 'it is not possible to give more than a vague idea to the outside world'; indeed, an important part of Victorian culture involved bringing the sciences to the widest possible audience. Salmon's phrase tells us that he already understood that there was now no point of contact between the work of a pure mathematician and anything that a layperson might find familiar, even an educated layperson who read *Nature*. However, Salmon then explained why, notwithstanding this, Cayley and his work should be celebrated:

> My subject is the life of a great artist who has had courage to despise the allurements of avarice or ambition, and has found more happiness from a life devoted to the contemplation of beauty and truth than if he had striven to make himself richer, or otherwise push himself on in the world. We do not classify

painters according to the numbers capable of appreciating their respective productions. On the contrary, we can understand that it is often the lowest style of art which will attract round it the largest circle of admirers. So the fact that it is a very limited circle which is capable of appreciating the beauty of the work done by a great mathematician should not prevent men from understanding that it is like the work done by a poet or painter, work done entirely for its own sake, and capable of affording lively pleasure both to the worker himself and his admirers, without any thought of material benefit to be derived from it.[14]

Salmon was here driven to adopt a new argument to justify the study of pure mathematics. Having admitted that to explain its nature was well-nigh impossible, he nevertheless had to find some way of obtaining for Cayley public honour and status. He achieved this by characterising the pure mathematician as an artist, a maker and contemplator of the beautiful, and thereby allied him not with scientists – an argument that, as we have seen, was no longer tenable – but with poets and painters. Airy had complained that pure mathematicians 'shut themselves up in a cavern', and thereby betrayed their scientific heritage; Salmon argued that, as artists, this is exactly what one would expect of them, and that they should be appreciated for the beauty, rather than the usefulness, of their productions. However, a mathematical theorem can be followed only after years of study, is not accessible to the layperson, and does not appear to be remotely 'like the work done by a poet or painter' at all. Yet Salmon, in drawing precisely that parallel, clearly believed that he was tapping into some public sentiment that made his characterisation of the pure mathematician plausible.

Salmon was writing at a time when there were well-defined views as to what art should be, in terms of which a page of mathematics by Cayley could not be the work of an artist because it was of the wrong form. It may have had a certain appeal to the non-mathematician in its organisation and symmetry, but that is of a kind that could be supplied by a designer or typographer, and had no connection with the meaning of the symbols. Hence Salmon's central claim, which was based on the purported beauty and truth that the symbolism represented, appeared paradoxical in conventional terms. However, Pater's characterisation of art as that which gives aesthetic pleasure to the beholder considerably widened the scope of what could be accounted beautiful. He referred only to 'the fairer forms of nature and human life', but nevertheless argued that it was the response that was significant, rather than the character of whatever evoked that response. If a page of mathematics gave the same aesthetic pleasure as a page of poetry, then there was nothing in his theory of aesthetics to deny that mathematics was beautiful, and that the pure mathematician, who created something beautiful, should be regarded as an artist.

Nevertheless, an apparently fundamental objection to describing the pure mathematician as an artist was that the work of an artist could be appreciated by everyone. Of course, opinions might vary as to the merits of a work of art, and a period of education and training might be required to understand the finer points, but there was nothing inherently inaccessible in the work of a poet or painter. Pure mathematics was quite different, and only a few individuals were so familiar with it that they could appreciate in aesthetic terms mathematical papers such as those of Cayley. All this Salmon readily conceded – he could hardly do otherwise – but once again the doctrines of aestheticism rendered the objection far less formidable than might at first appear, because the aesthetes believed themselves to have higher powers of perception than the rest of the population, and were thereby enabled to take rapturous pleasure in things that normal people regarded with indifference. Thus the aesthetes constituted a self-defining elite, and recognition that pure mathematicians – the only individuals to whom mathematical ideas and demonstrations were accessible – also constituted an elite was not fatal to Salmon's claims. Quite the contrary: his argument – that the value of pure mathematics lay in the pleasure that it gave to a very limited circle of pure mathematicians – paralleled the arguments made on their own behalf by the aesthetes.

The final element in Salmon's presentation of the pure mathematician was a claim that such a life had been well-spent. Salmon put before his readers not only that Cayley should be regarded as a great artist, but also that his life had been 'devoted to the contemplation of beauty and truth', and therefore that he should be revered, having forgone wealth and advancement. Support for the argument that a pure mathematician had spent his time in the best possible way, even though he had produced nothing of value to anyone other than himself and a few admirers, is once again found in Pater's *Studies*:

> We have an interval, and then our place knows us no more. Some spend this interval in listlessness, some in high passions,

the wisest in art and song. For our one chance is in expanding that interval, in getting as many pulsations as possible into the given time. High passions give one this quickened sense of life, ecstasy and sorrow of love, political or religious enthusiasm, or the 'enthusiasm of humanity.' Only, be sure it is passion, that it does yield you this fruit of a quickened, multiplied consciousness. Of this wisdom, the poetic passion, the desire of beauty, the love of art for art's sake has most; for art comes to you professing frankly to give nothing but the highest quality to your moments as they pass, and simply for those moments' sake.[15]

An existence governed solely by the desire to indulge personal pleasures – the kind of existence with which Airy had taxed Cayley – was here elevated to a high purpose in respect of which no defence or apology was needed; and so in Salmon's characterisation the life of a pure mathematician was well-spent because he was maximising the exposure of himself and his like-minded followers to 'lively pleasure'.

Pater's *Studies* therefore provided not only controversial views for the aesthetes to adopt and enthusiastically embellish, but also the intellectual framework that enabled Salmon to introduce the idea of the pure mathematician as an artist; and there was no question that artists, the creators of beauty, were valuable members of any civilised society. Salmon was not saying that pure mathematicians *were* aesthetes, which would clearly have been untrue, but his argument was plausible because he presented their relationship with pure mathematics as being equivalent to the aesthetes' relationship with art.

Salmon was one of Britain's foremost geometers, a priest and Regius Professor of Divinity at Trinity College, Dublin; in addition, he was the author of many theological works, which substantially outnumbered his mathematical publications. On the other hand, Pater's aesthetic creed was not underpinned by any form of religious doctrine, because it promoted a way of life in which the most important elements were the pleasurable experiences of the individual. Consequently it is unlikely that there was any conscious intention on Salmon's part to appeal directly to the philosophy of the aesthetes in order to bolster support for pure mathematicians; but notwithstanding this, he clearly used arguments that the aesthetes had deployed. The parallels are set out in Table 7.1 below, which compares the claims of Pater and the aesthetes with those made by Salmon, and shows that Salmon was

Table 7.1 The creed of the aesthetes compared with Salmon's characterisation of pure mathematics.

Pater and the aesthetes	George Salmon
Beauty is not restricted to the fine arts	The world of pure mathematics is beautiful
Only an elite has the higher powers of perception necessary for the true appreciation of beauty	Only a trained elite can perceive the beauty of pure mathematics
The value of a beautiful thing lies in the pleasure that it gives to those who can appreciate it	The value of pure mathematics lies in the pleasure that it gives to those who can appreciate it
To give oneself over to experiencing such pleasure is to spend one's life well	The life of a pure mathematician has been well spent in 'lively pleasure'

formulating his argument in a manner already familiar to his readers, and that could therefore be subsumed under the terms of a much wider cultural debate.

When Cayley came to give his presidential address to the British Association's conference, he painted a picture of the pure mathematician as a wanderer in a wonderful new world. Although an explorer, he was not an explorer in the manner of Du Chaillu, out to make a scientific evaluation of what he saw, or to capture specimens; instead, his approach was far more bucolic, and followed Salmon by invoking the beauty of pure mathematics. This was a gift to *Punch*, which predictably made the most of the opportunity by first quoting a particularly lyrical passage:

> It is difficult to give an idea of the vast extent of modern Mathematics. This word, 'extent', is not the right one; I mean extent crowded with beautiful detail – not an extent of mere uniformity, such as an objectless plain, but of a tract of beautiful country seen at first in the distance, but which will bear to be rambled through and studied in every detail of hillside and valley, stream, rock, wood, and flower. But as for anything else, so for a mathematical theory, beauty can be perceived but not explained.

Punch then appended a long and exceedingly tiresome verse entitled 'The Song of a Sciolist at Southport', full of mathematical references and laboured humour that showed the extent to which pure mathematicians and their work had for once been under the public's gaze. After the quotation from Cayley's address, we find:

> Talk not of mountains,
> Of streams and fountains,
> For what *is* land or water, and *what* is wood,
> To contemplations
> Of sweet equations
> As seen by CAYLEY, or known to SPOTTISWOODE?

This set the tone; the next few lines introduce Oscar Wilde as a poet, non-Euclidean space, four-dimensional space, and $X + iY$, before reaching 'least factors':

> Though Glaisher – rum thing! –
> Has been doing *something*
> With the missing three out of the first nine millions!
> Oh, blissful duty
> To explore the beauty
> Of elliptic and multiple *theta* functions!

The reference to 'Glaisher' cites work done by James Glaisher senior in preparing factor tables covering the integers between 3,000,000 and 6,000,000, which had been the uncalculated three million out of the first nine; the final volume, for the sixth million, had just been published. The second half of the verse uses the aesthetically charged words 'blissful' and 'beauty' to describe functions that were of especial interest to his son James W L Glaisher, and that were also mentioned extensively by Cayley in his address. And then:

> The mathematics
> Must inspire ecstatics
> Which should thrill an Æsthete with 'intense' compunctions.
> *But* you 'can't explain it!'[16]

With these lines, *Punch* extracted humour by trading on the ostentatious behaviour of the aesthetes, as du Maurier had done with Jellaby Postlethwaite and other creations, and as Gilbert had done in *Patience*. This time, however, there is an explicit connection made between the aesthetes and pure mathematicians, which shows that their respective claims had a relationship that readers of *Punch* were expected to understand, and that would have had no cogency had it not existed at a deeper and more sober level, once the humour and satire were stripped away.

Salmon's portrait of Arthur Cayley was a public recognition that in Britain, where the tradition of valuing pure mathematics for its role in natural philosophy was so strong, a new approach was at last necessary, for the old arguments had run their course, bore no relation to the manner in which pure mathematics was practiced, and were unacceptable to the pure mathematicians themselves. This new approach aligned itself much more closely to pure mathematicians' true motivations, and so gave them greater confidence when talking about and explaining their discipline. Unflattering references to pure mathematics became increasingly rare, and pure mathematicians had no reason to be defensive, for when it suited them they could now present themselves in the role of creators of beauty. The parallels between their claims and those of the aesthetes also encouraged the idea of the pure mathematician as an inhabitant of a rarefied intellectual realm that had little contact with everyday, mundane affairs; and it was this portrayal that would eventually reverse the relative degrees of esteem in which pure and applied mathematicians were held. Consequently, by the turn of the century pure mathematics in Britain had gained a permanent place alongside other disciplines that were deemed to be worthy of a lifetime's study.

Nevertheless, these changes did not happen overnight, and it was inevitable that some of the older generation continued to uphold the 'servant' tradition of pure mathematics that had obtained in their youth. Such a one was William Henry Besant, the noted coach, who was born in 1828, was 20 years senior to Glaisher, and had been senior wrangler in 1850. The reader may recall that in 1890, when the Senate at Cambridge was debating proposed changes to the Mathematical Tripos, he made some remarks to the effect that the value of pure mathematics lay only in its utility.[b] Glaisher's reply included the hope that 'Pure Mathematics were no longer regarded as merely weapons and tools for dealing with physical questions', and he did not feel the need to gloss the discipline so that it appeared to possess scientific attributes, as Sylvester had done 20 years earlier. Yet Glaisher was ambivalent concerning pure mathematics: on the one hand, he insisted that it was

[b] See p. 123.

worthy to be studied in its own right without any reference to actual or potential utility; but on the other hand, he was unwilling, or unable, to explain *why* its value was nevertheless equal to that of the applied sciences.

A greater degree of confidence was apparent in 1897 when Andrew Russell Forsyth, who was that year's president of Section A (mathematical and physical science) of the British Association, delivered at their conference an address explaining the nature of pure mathematical activity.[17] There must have been a suggestion that pure mathematics was by then so abstract that it should no longer be discussed in Section A, because Forsyth began by setting out a defence of its continued inclusion. He did not allow himself to dwell on the aesthetic qualities of pure mathematics, since such an argument would not have found favour with an audience drawn mainly from applied mathematicians and scientists; but as a pure mathematician he also did not wish only to stress the uncontested point that pure mathematics had proved to be of great benefit in the sciences. After noting the passing of the old view of pure mathematics, he made some comments in which we can detect elements of the aesthetes' opinion of themselves:

> Feelings of hostility, if ever they were widely held, have given way to other feelings, which in the mildest form suggest toleration and acquiescence, and in the most extreme form suggest solemn respect and distant wonder. By common consent, we are allowed without reproach to pursue our aims; though those aims sometimes attract but little sympathy. It is not so long since, during one of the meetings of the Association, one of the leading English newspapers briefly described a sitting of this Section in the words, 'Saturday morning was devoted to pure mathematics, and so there was nothing of any general interest': still, such toleration is better than undisguised and ill-informed hostility. But the attitude of respect, I might almost say of reverence, is even more trying: we mathematicians are supposed to be of a different mould, to live far up the heights above the driving gales of controversy, breathing a rarer intellectual atmosphere, serene in impenetrable calm. It is difficult for us to maintain the gravity of demeanour proper to such superior persons; and perhaps it is best to confess at once that we are of the earth, earthy, that we have our differences of

opinion and of judgement, and that we can even commit the Machiavellian crime of making blunders.

Unlike Henry Smith in 1876 and Samuel Roberts in 1882, Forsyth had the confidence to admit in a light-hearted fashion the fallibility of pure mathematicians; the defensiveness that was evident a generation earlier was on the wane. Then, after echoing Salmon in asserting the impossibility of explaining even in the most general terms what pure mathematics was about, he advised his audience that although his address would show how beneficial the work of pure mathematicians had been to other branches of science, he would also:

> attempt ... to make a bold claim that the unrestricted cultivation of pure mathematics is desirable in itself and for its own sake. Some – I should like to believe many – who are here will concede this claim to the fullest extent and without reservation; but I doubt whether this is so in general. And yet the claim is one that needs to be made before an English-speaking audience.

He was making the point that it would be unnecessary to make such a claim before a Continental audience, where the value of pure mathematics as an abstract study had long been accepted. After acknowledging that the British attitude towards new pure mathematics had been one of 'apathy rather than attention', he continued:

> Now it is not enough for my purpose to be told that the British Association includes all science in its scope, and consequently includes pure mathematics. A statement thus made might be framed in a spirit of mere sufferance; what I wish to secure is a recognition of the subject as one which, being full of life and overflowing with the power of growth, is worthy of the most absorbing devotion.

He then referred to attempts that were still being made to limit the purview of pure mathematics, and said that, on the contrary, other branches of science would benefit to the greatest extent if pure mathematicians were left to their own devices, doing precisely what they wanted. The clear message in this part of his address was that pure mathematics had come of age as an independent discipline that should be pursued for its own sake, irrespective of any impact that it might have elsewhere. History showed that other disciplines did often benefit from advances in pure mathematics, but although critics should be satisfied by this, it quite rightly formed no part of the motivation of pure mathematicians, nor should it be used to impose boundaries on their research.

Salmon's sketch of Arthur Cayley in 1883 had assumed that his name and work would be unfamiliar to most readers. However, by the turn of the century Cayley's name, at least, had become almost synonymous for 'mathematician' in the public's mind, as we can understand from a series of short stories that the science populariser Grant Allen published in the *Strand Magazine* in 1898–1899, some four years after Cayley's death. Allen was a prolific writer on many subjects whose work on natural history and evolution had been praised by Darwin and Huxley, and whose friend he was;[18] although a 'popular' writer, he was 'integrated into an intellectually respectable social circle'.[19] When interviewed, he said that what he enjoyed best was writing 'a popular story for popular consumption, introducing a good deal of the scientific element in an unobtrusive form'; he aimed his work to 'hit the taste of a wide body of readers'.[20]

Allen's stories in the *Strand* concerned a Girton Girl[c] who had achieved high honours in mathematics, and whose capacity for logical, unemotional and disinterested reasoning – clearly the hallmark of a pure mathematician – led her safely through several scrapes and adventures. To achieve the characterisation that he desired, he tapped into the common belief that Girton Girls in general had a particular fondness for mathematics, a belief that had even been expressed poetically – each verse of Andrew Lang's *Ballade of the Girton Girl*, dating from 1885, ends with: 'But her forte's to evaluate π.'[21] By this means Allen was able to avoid giving much overt description, and direct references to his heroine's mathematical background appear rarely. However, to put beyond doubt her ability to 'evaluate π', he gave her the name Lois Cayley; as an experienced author in touch with his public he knew that this would act throughout the stories as a reminder of the mathematical cast of mind believed to be natural for a Girton Girl (see Figure 7.2). The stories were evidently popular, for under the title *Miss Cayley's Adventures* they were republished three times as a book – twice in London and once in New York.[22]

---

[c] A student or former student at Girton College, Cambridge, which was then for women only.

Figure 7.2 Lois Cayley: clearly a mathematician.

By the turn of the century the idea that pure mathematics was a discipline quite independent of the applied sciences had been accepted not only by a new generation of mathematicians but also by many non-mathematicians whom pure mathematics in some way touched. For example, in 1901 the cataloguers of the works on pure mathematics in the library of Newcastle-upon-Tyne set out their principles of selection:

> The books and tracts which are included in this catalogue are those in which the first object of the writer has been the exposition of pure mathematics ... unless the main object of the writings has seemed to be the upholding of purely mathematical rather than physical truth, it has been thought better to assign them to fields of knowledge which they most directly and immediately serve. ... In such writings as have been rejected, Number has been made subservient and has not been investigated purely for its own sake.[23]

This was a formalisation of the breach with natural philosophy that had been widening throughout the second half of the nineteenth century, for the earlier characterisation of pure or abstract mathematics was precisely to 'serve' other fields of knowledge.

In 1907 Bertrand Russell used the concept of pure mathematics as a thing of beauty in an attempt to bring an understanding of the mindset of the pure mathematician to a wider audience. He was writing in *The New Quarterly*, a journal that offered popular but informed articles on the humanities and the sciences, and so was aimed at a readership with a correspondingly wide range of interests. In the first volume, which contained Russell's article, offerings of a scientific nature by Lord Rayleigh, his son R J Strutt, Oliver Lodge and Norman Campbell are found alongside others dealing with literature and the arts by Lytton Strachey, Arthur Symons, Hilaire Belloc, Max Beerbohm, H G Wells and G K Chesterton; the editor was Desmond MacCarthy, the author and literary critic. Russell wanted to impress his views upon the world at large, and had no desire to write for specialists such as the members of the London Mathematical Society, or even for scientifically inclined laypeople such as comprised the majority of the British Association, or read *Nature*.

Although Russell is now remembered as a philosopher, he had been joint seventh wrangler in 1893 and regarded himself primarily as a mathematician until shortly before the First World War, saying that 'At the age of eleven, I began Euclid ... From that moment until Whitehead and I finished *Principia Mathematica*, when I was thirty-eight, mathematics was my chief interest, and my chief source of happiness.'[24] Russell believed that pure mathematics, properly understood, was a completely new discipline grounded in the symbolic logic of Boole, and defined it as 'the class of all propositions of the form "p implies q", where p and q are propositions containing one or more variables, the same in the two propositions, and neither p nor q contains any constants except logical constants'.[25] It is therefore unsurprising that he presented pure mathematics to the readers of *The New Quarterly* as a somewhat ascetic study, but one which nevertheless could provoke an aesthetic response of the kind described by Pater. Russell wrote:

> Mathematics, rightly viewed, possesses not only truth, but supreme beauty – a beauty cold and austere, like that of sculpture, without appeal to any part of our weaker nature, without the gorgeous trappings of painting or music, yet sublimely pure, and capable of a stern perfection such as only the greatest art can show. The true spirit of delight, the exaltation, the sense of being more than man, which is the touchstone of the highest excellence, is to be found in mathematics as surely as in poetry.[26]

In this panegyric to pure mathematics Russell went considerably further than did Salmon in 1883, for Cayley's working methods were not always the most refined. Salmon drew attention to this, as opposed to the sly and underhand methods of mathematicians (Continental, of course) who adopted non-constructive proofs, when he said:

> Men weak in power of calculation have often exhibited beautiful exercises in ingenuity in their attempts to arrive at results by some shorter process. Such a master of algebra in all its forms as Cayley was not to be dismayed by any amount of calculation, and he therefore has been able to trample down many a difficulty which an inferior in this respect might have evaded by some ingenious oblique method.[27]

British mathematicians had long taken pride in their ability to calculate, and regarded it as a matter of honour to be able to out-perform their Continental neighbours; this had been the principal motive for William Shanks's first record-breaking computation of π to 607 places of decimals in 1853.[28] Salmon had been affirming Cayley's place in the pantheon of mental calculators by highlighting the blood, sweat and tears that were sometimes necessary for the production of beauty, whereas Russell's portrayal of the work of the pure mathematician was comparatively insipid.

Nevertheless, Russell agreed with Salmon that the pure mathematician obtained his fulfilment from the production of beauty, and then added that the link between pure mathematics and the natural world had finally been severed, introducing again the tropes of the mathematician as creative artist and explorer that are already familiar through Salmon and Sylvester:

> The objects considered by mathematicians have, in the past, been mainly of a kind suggested by phenomena: but from such restrictions the abstract imagination should be wholly free. A reciprocal liberty must thus be accorded: reason cannot dictate to the world of facts, but the facts cannot restrict reason's privilege of dealing with whatever objects its love of beauty may cause to seem worthy of consideration. Here, as elsewhere, we build up our own ideals out of the fragments to be found in the world; and in the end it is hard to say whether the result is a creation or a discovery.

He then urged that students should be inculcated with a feeling for the beauty of a mathematical demonstration, a procedure that he introduced as a pedagogic novelty:

> It is very desirable, in instruction, not merely to persuade the student of the accuracy of important theorems, but to persuade him in the way which itself has, of all possible ways, the most beauty. The true interest of a demonstration is not, as traditional modes of exposition suggest, concentrated wholly in the result; where this does occur, it must be viewed as a defect, to be remedied, if possible, by so generalising the steps of the proof that each becomes important in and for itself. An argument which serves only to prove a conclusion is like a story subordinated to some moral which it is meant to teach: for aesthetic perfection, no part of the whole should be merely a means.

Finally, in his concluding passage he offered sentiments that are very close to those that Hardy was to promulgate more than a decade later in his inaugural at Oxford; it seems likely that if Russell had remained a mathematician, and not reinvented himself as one of the twentieth century's most distinguished philosophers, then he rather than Hardy would be consistently quoted today to express pure mathematicians' attitude towards their discipline. Russell wrote:

> In a world so full of evil and suffering, retirement into the cloister of contemplation, to the enjoyment of delights which, however noble, must always be for the few only, cannot but appear as a somewhat selfish refusal to share the burden imposed upon others by accidents in which justice plays no part. Have any of us the right, we ask, to withdraw from present evils, to leave our fellow-men unaided, while we live a life which, though arduous and austere, is yet plainly good in its own nature? When these questions arise, the true answer is, no doubt, that some must keep alive the sacred fire, some must preserve, in every generation, the haunting vision which shadows forth the goal of so much striving. But when, as must sometimes occur, this answer seems too cold, when we are almost maddened by the spectacle of sorrows to which we bring no help, then we may reflect that indirectly the mathematician often does more for human happiness than any of his

more practically active contemporaries. The history of science abundantly proves that a body of abstract propositions – even if, as in the case of conic sections, it remains two thousand years without effect on daily life – may yet, at any moment, be used to cause a revolution in the habitual thoughts and occupations of every citizen. The use of steam and electricity – to take striking instances – is rendered possible only by mathematics. In the results of abstract thought, the world possesses a capital, of which the employment in enriching the common round has no hitherto discoverable limits. Nor does experience give any means of deciding what parts of mathematics will be found useful. Utility, therefore, can be only a consolation in moments of discouragement, not a guide in directing our studies.[29]

Russell was here repeating the views of Forsyth: it is a fact that pure mathematics enabled advances other disciplines, but if modern pure mathematics does so then it is fortuitous; such a prospect does not, and cannot, motivate pure mathematicians, whose purpose is the production and contemplation of beauty. By characterising pure mathematicians in that way, by denying the value of empirical considerations, and by using the language of symbolic logic to define pure mathematics in a manner that few applied mathematicians would recognise, Russell was effectively raising the question of what, if anything, pure and applied mathematicians had in common, and in what sense, apart from using a common symbolism, they belonged to the same discipline.

Sir George Darwin addressed this question far more explicitly when giving the opening address to the 1912 International Congress of Mathematicians at Cambridge, which was hosted by the Cambridge Philosophical Society, of which he was president.[30] He had been a member of the London Mathematical Society since 1868 and was a distinguished applied mathematician, but he began with an admission:

> The science of mathematics is now so wide and is already so much specialized that it may be doubted whether there exists today any man fully competent to understand mathematical research in all its many diverse branches. I, at least, feel how profoundly ill equipped I am to represent our [Cambridge Philosophical] society as regards all that vast field of knowledge which we classify as pure mathematics. I must tell you frankly

that when I gaze on some of the papers written by men in this room I feel myself much in the same position as if they were written in Sanskrit.

He then drew attention to how few pure mathematicians of note Cambridge had recently produced – namely Cayley and Sylvester – when compared to the long roll call of applied mathematicians and mathematical physicists who had international reputations. Notwithstanding this, he suggested that achievement in pure mathematics is easier, in the sense that any advance is an advance, whereas in applied mathematics a mathematical advance is insufficient in isolation; to succeed, it must also relate appropriately to the physical world. A failure in the latter consideration means that the mathematics as a whole also fails, notwithstanding the logical correctness of the mathematical argument. As an example he gave Lord Kelvin's investigations concerning the age of the Earth. Kelvin's final calculations were made shortly before the phenomenon of radioactivity was discovered; thus his mathematical model of the cooling Earth was incomplete in a vital respect, and so understated the maximum age by a very great margin. However, there was nothing wrong with his mathematics as such, but considered as a mathematical model it failed to reflect correctly the physical conditions, and so was almost immediately superseded.

Darwin did not make these observations simply because of their intrinsic interest, for he clearly believed that applied mathematicians had come to be thought of as inferior to pure, and he wanted to draw attention to the restrictions by which they were governed. The apparent superiority of pure mathematicians arose, he suggested, out of the deemed aesthetic qualities of their demonstrations, but applied mathematicians worked under circumstances that limited their ability to achieve such perfection. He therefore concluded, somewhat apologetically:

> I appeal then for mercy to the applied mathematician and would ask you to consider in a kindly spirit the difficulties under which he labors. If our methods are often wanting in elegance and do but little to satisfy that esthetic sense of which I spoke before, yet they are honest attempts to unravel the secrets of the universe in which we live.

Such comments confirm the extent to which the former view – that the pure mathematician's only role was to provide an essential ingredient of mixed mathematics – had become nothing more than an historical curiosity, and bore no relation to the manner in which mathematicians thought about or practised mathematics at the beginning of the twentieth century.

During the century's first decade the distinctive natures of the two disciplines was given further prominence by the arrival of Hardy, the militant pure mathematician who was the rising star of British mathematics and was soon to dominate the London Mathematical Society. He believed that the pure mathematician's principal task was to undertake research, and the strength of his commitment became evident in 1919 when he moved from Cambridge to Oxford as the new Savilian Professor of Geometry. This was not the branch of mathematics in which he had made his reputation, but that was no obstacle to his appointment because in earlier times 'geometry' often stood proxy for 'mathematics'; and for Oxford, a mathematical backwater for more than a century, his appointment represented a substantial coup. He was by then the best-known pure mathematician in Britain, not just on account of his mathematical powers, which were indeed very great, but also because of his charisma and his championing of the Indian mathematician Srinivasa Ramanujan, who had recently captured the public imagination.

Although Hardy had been only fourth wrangler in 1898 he was first Smith's prizeman in 1901, and thereafter his career at Trinity College was one of unalloyed success. In a succession of journal articles, book reviews and mathematical papers he brought to British mathematics an uncompromising belief in the need for logical rigour at every step of a mathematical demonstration, and in 1908 his famous textbook *Pure Mathematics* had shown how such principles could be introduced at an elementary undergraduate or even upper-sixth-form level. Hardy's delight in rigour was later illustrated by Russell's anecdote in which Hardy 'told me once that if he could find a proof that I was going to die in five minutes he would of course be sorry to lose me, but this sorrow would be quite outweighed by pleasure in the proof. I entirely sympathized with him and was not at all offended.'[31] In his own particular field of analysis (and *Pure Mathematics* is essentially a textbook of analysis) Hardy established with Littlewood, also of Trinity, a partnership unique in the annals of mathematics for its fruitfulness: of Hardy's three-hundred-odd papers, about one hundred were co-authored with Littlewood.

Hardy agreed with the opinion that Littlewood was an even stronger mathematician, and Littlewood himself said that 'My standard role in a joint paper was to make the logical skeleton, in shorthand – no distinction between r and r^2, 2π and 1 etc., etc. But when I said "Lemma 17", it stayed Lemma 17.'[32] Nevertheless, it was Hardy who was in the public eye: in 1920 the German number theorist Edmund Landau remarked that 'the mathematician Hardy-Littlewood was the best in the world, with Littlewood the more original genius and Hardy the better journalist'.[33] Indeed, so contrasted were they in their attitude towards publicity that there were various pleasantries being bandied about posited on the non-existence of Littlewood, and Landau solemnly assured the Göttingen Mathematical Society that 'they were two quite different men, for he had seen them both together'.[34] Harald Cramér, who had taken a photograph of Hardy and Littlewood standing side by side, later wrote to Albert Ingham: 'At that time there was a theory current among certain mathematicians that Littlewood did not exist, and my photo was regarded as an existence proof.'[35] The theory presented Littlewood as a fictional character invented by Hardy in order to have someone to blame for any errors that were detected in his (or their) papers.[36] Almost certainly Landau had been one of the 'certain mathematicians' before seeing Hardy and Littlewood together (see Figure 7.3).[37]

After the First World War, with an unassailable public and academic reputation, and seemingly tailor-made for a career as a resident bachelor don at Trinity, Hardy was clearly the heir to the Sadleirian Chair, of which the incumbent was Ernest Hobson; and following Hobson's resignation in 1931 Hardy did indeed succeed him. Yet when, by his own estimation, his mathematical ability was approaching its peak,[38] we find him accepting a chair at Oxford, a university with virtually no tradition or reputation in mathematics, and where neither his colleagues nor his students would provide him with the stimulus that he was accustomed to receiving at Cambridge. If this was a career move then it was certainly a puzzling one, but we know of two reasons. The first arose in 1916, when Bertrand Russell, then a lecturer at Trinity, had been fined for prejudicing the recruitment and discipline of the armed forces and had consequently lost his lectureship. Hardy had been one of the Trinity dons who rallied to his defence and, after the War, called unsuccessfully for his reinstatement. Feelings ran high, and Hardy suffered considerable unhappiness. More than 20 years later, in 1942, he

Figure 7.3 The existence proof: Hardy on the left, Littlewood on the right.

was still working for Russell's return, and thought the principles sufficiently important to have an account of the controversy privately printed.[39] In this he made clear the depth of his feeling, saying: 'I felt bitterly about the matter at the time, and feel strongly about it still, and I have not changed my views on any important point.'[40] In particular, his life at that time had been 'definitely unpleasant', which was 'an important factor in my own decision to try to move to Oxford'.[41] However, the principal reason for his move he set out at the time in a letter to the Master of Trinity, J J Thomson. He said that his responsibilities at Cambridge were such that there was a 'risk of wearing myself out prematurely', and that he found himself with virtually no time for 'the researches which are the principal permanent happiness of my life'. He acknowledged that he was taking what would seem to be a retrograde step, and that his students would not be 'in any way comparable, either in numbers or in quality, to those I have taught here [at Cambridge]'; yet he considered it a worthwhile move so as to secure 'more leisure' in which to pursue his true vocation.[42]

A few months after his appointment he gave his inaugural lecture as Savilian Professor of Geometry before a distinguished audience: he did not, of course, talk on geometry, but on one of his particular interests, namely Waring's problem, which deals with the expression of integers as sums of powers of other integers. However, he also included opinions on the subject of pure mathematics generally; he was never one to say either more or less than exactly what he meant. At Cambridge he had been an enthusiastic advocate of the mathematical logic of Whitehead and Russell, and in his inaugural he made his support clear with some typically trenchant remarks:

> In the sphere of philosophy we mathematicians put forward a strictly limited but absolutely definite claim ... that there are a number of puzzles, of an abstract and elusive kind, with which the philosophers of the past have struggled ineffectually, and of which we can now give a quite definite and explicit solution. There is a certain region of philosophical territory which it is our intention to annex. It is a strictly demarcated region, but it has suffered under the misrule of philosophers for generations, and it is ours by right; we propose to accept the mandate for it, and to offer it the opportunity for self-determination under the mathematical flag.[43]

This was Hardy in an optimistic and confident mood, yet at the beginning of his address there is a short passage that would have sounded strange to those who were expecting to hear a panegyric on pure mathematics and its practitioners:

> For my own part, I trust that I am not lacking in respect either for my subject or myself. But, if I am asked to explain how, and why, the solution of the problems that occupy the best energies of my life is of importance in the general life of the community, I must decline the unequal contest: I have not the effrontery to develop a thesis so palpably untrue. ... A pure mathematician must leave to happier colleagues the great task of alleviating the

sufferings of humanity. I suppose that every mathematician is sometimes depressed, as certainly I often am myself, by this feeling of helplessness and futility. I do not profess to have any very satisfactory consolation to offer. It is possible that the life of a mathematician is one which no perfectly reasonable man would elect to live.[44]

Whereas Russell had somewhat despairingly introduced an argument from unspecified but potential utility as a means of mitigating the pure mathematician's distress when facing the troubles of the world, Hardy did not even allow himself that consolation; but he was far more ascetic as a person than the libidinous Russell. However, Hardy was perhaps aware that his comments would not arouse universal approbation, for Littlewood records that Hardy let him have a copy of the lecture 'only on the condition that we were never to mention it'.[45] Hardy's was the voice of the unadulterated pure mathematician, melancholy but uncompromising in disclaiming the need for any external validation. He was not the first to utter such sentiments, but never before had they been offered by such an esteemed and high-profile British mathematician, and on such a public occasion.

At the end of his career Hardy was to ensure even wider exposure for his views through his famous little book *A Mathematician's Apology*, which gained a new generation of readers when it was reissued in 1967 with a long introduction by C P Snow.[46] For more than 70 years this book has helped to shape the public view of pure mathematicians and pure mathematics, yet it was in his 1920 inaugural that he made explicit, with his usual directness, the moral dilemma that pure mathematicians faced. He had been profoundly affected by the First World War, which he believed had brought with it the perversion of scientific knowledge for ignoble ends, whereas pure mathematics, although it could do no good in the world, at least could do no harm. Furthermore, he believed that his life's work lay in mathematical research, notwithstanding that he very much enjoyed lecturing, at which he was extremely good:[47] the American Norbert Wiener, then a teenager studying with Russell, later wrote that 'In all my years of listening to lectures in mathematics, I have never heard the equal of Hardy for clarity, for interest, or for intellectual power.'[48] Nevertheless, Hardy would not allow his research in pure mathematics to be blighted, as Henry Smith's had been, by the pedagogical and administrative demands of college and university. As a lecturer or tutor at Cambridge he would never have enjoyed the freedom that he acquired as a professor at Oxford, and his decision to move there made the 1920s the most productive decade of his career. He was the first British pure mathematician of international standing to promote pure mathematics as a study justified by its aesthetics, and his success at Oxford meant that there was now no turning back – the days of pure mathematics as 'the servant of the sciences' were finally over. As he was to write of mathematics in the *Apology*: 'Beauty is the first test: there is no permanent place in the world for ugly mathematics.'[49]

Notes and References

1. See the advertisement opposite the title page in the second edition of Pater's *Appreciations, with an Essay on Style* (London: Macmillan, 1890).
2. Walter H Pater, *Studies in the History of the Renaissance* (London: Macmillan, 1873), pp. vii–ix, 213.
3. See the footnote that Pater added to the Conclusion when it was restored to favour in subsequent editions.
4. Ian Small, *Conditions for Criticism: Authority, Knowledge, and Literature in the Late Nineteenth Century* (Oxford: Clarendon Press, 1991), p. 6.
5. Small, *Conditions for Criticism*, p. 7.
6. Small, *Conditions for Criticism*, p. 6.
7. H Montgomery Hyde (ed.), *The Trials of Oscar Wilde* (London: Hodge, 1948), pp. 52, 150.
8. R V Johnson, *Aestheticism* (London: Methuen, 1969), p. 1.
9. Small, *Conditions for Criticism*, p. 128.
10. Henry A Lytton, *The Secrets of a Savoyard* (London: Jarrolds, 1922), p. 190.
11. Lytton, *Secrets*, p. 134.
12. M G Brock and M C Curthoys, *The History of the University of Oxford, Volume VII: Nineteenth-Century Oxford*, 2 (Oxford: Clarendon Press, 2000), p. 452.
13. George Salmon, 'Arthur Cayley', *Nature*, 28, 725 (1883), 481–485, at 481.
14. Salmon, 'Arthur Cayley', 483–484.
15. Pater, *Studies*, pp. 212–213.
16. *Punch, or the London Charivari*, 85 (29 September 1883), 154.
17. A R Forsyth, 'Presidential Address to Section A of the British Association', *Report of the Sixty-seventh Meeting of the British Association for the Advancement of Science* [1897] (London: John Murray, 1898), pp. 541–549.
18. See letters to Allen from Darwin and Huxley, quoted in Edward Clodd, *Grant Allen: A Memoir* (London: Grant Richards, 1900), pp. 111, 112.
19. Kate Jackson, *George Newnes and the New Journalism in Britain, 1880–1910: Culture and Profit* (Aldershot: Ashgate, 2001), p. 105.
20. Quotations from an interview with Allen in *The Million*, 2, 36 (1892), 129.
21. Andrew Lang, *Rhymes a la Mode*, fourth edition (London: Kegan Paul, 1890), pp. 43–44.
22. The image is taken from Grant Allen, *Miss Cayley's Adventures* (New York: Putnam's, 1899), p. 236.

23. Basil Anderton and R T Richardson (comps.), 'Preface', *Catalogue of the Books and Tracts of Pure Mathematics in the Central Library* [of Newcastle-upon-Tyne] (Newcastle-upon-Tyne: Andrew Reid, 1901).
24. Bertrand Russell, *The Autobiography of Bertrand Russell: 1872–1914*, I (London: Allen & Unwin, 1967), p. 36.
25. Bertrand Russell, *The Principles of Mathematics* (Cambridge: Cambridge University Press, 1903), p. 3.
26. Bertrand Russell, 'The Study of Mathematics', *The New Quarterly*, 1, 1 (1907), 31–44, at 33.
27. Salmon, 'Arthur Cayley', 483.
28. Andrew Warwick, 'The Laboratory of Theory, or, What's Exact about the Exact Sciences?', in M Norton Wise (ed.), *The Values of Precision* (Princeton, NJ: Princeton University Press, 1995), p. 318.
29. Quotations from Russell, 'Study', 41–44.
30. Darwin's address can be found in Virgil Snyder's account of the Congress: *Bulletin of the American Mathematical Society*, 19, 3 (1912), 107–130.
31. Bertrand Russell, 'Why I Took to Philosophy', in *Portraits from Memory, and Other Essays* (London: Allen & Unwin, 1956), p. 19.
32. Quoted in Béla Bollobás (ed.), *Littlewood's Miscellany* (Cambridge: Cambridge University Press, 1986), pp. 118–119.
33. Quoted in Bollobás, *Littlewood's Miscellany*, pp. 1–2.
34. A Ostrowski recalled the incident in his obituary speech for Hardy given to the Basle Mathematical Society on 28 January 1948; typescript copy in the Library of Trinity College, Cambridge, Add. Ms. b. 175.
35. Cramér's letter to Ingham dated 23 June 1963, in the Library of Trinity College, Cambridge, Add. Ms. b. 175.
36. C P Snow, 'Foreword' to G H Hardy, *A Mathematician's Apology* (Cambridge: Cambridge University Press, 1967), p. 29; originally published (without the foreword) 1940.
37. Norbert Wiener, *Ex-Prodigy: My Childhood and Youth* (New York: MIT Press, 1964), p. 24; originally published 1953.
38. Hardy, *Apology*, p. 148.
39. Paul Delany, 'Russell's Dismissal from Trinity: A Study in High Table Politics', *Russell: The Journal of the Bertrand Russell Archives*, ns 6, 1 (1986), 39–61, at 39.
40. G H Hardy, *Bertrand Russell and Trinity: A College Controversy of the Last War* ('Printed for the Author' at the University Press, Cambridge, 1942), p. 2.
41. Hardy, *Bertrand Russell and Trinity*, p. 10.
42. Letter from Hardy to Thomson dated 13 December [1919]: Cambridge University Library, reference MS Add. 7654/H26.
43. G H Hardy, *Some Famous Problems of the Theory of Numbers, and in particular Waring's Problem: An Inaugural Lecture delivered before the University of Oxford* (Oxford: Clarendon Press, 1920), pp. 5–6; he gave the lecture on 18 May 1920.
44. Hardy, *Some Famous Problems*, p. 4.
45. Typescript autobiographical notes by Littlewood in the papers of Arthur Lehman Goodhart, Bodleian Library, MSS.Eng.d.2384, fo. 94.
46. For a critical discussion of the book and of Snow's introduction, see L J Mordell, 'Hardy's "A Mathematician's Apology"', *The American Mathematical Monthly*, 77, 8 (October 1970), 831–836.
47. Hardy, *Apology*, p. 149.
48. Wiener, *Ex-Prodigy*, p. 190.
49. Hardy, *Apology*, p. 85.

EPILOGUE

From the time of Plato and Aristotle, more than 2,000 years ago, questions concerning the nature of mathematics and mathematical entities have engaged the attention of countless philosophers and metaphysicians, some of whom were among the finest minds of their generation. Nevertheless, the difficulties encountered in attempting to answer these questions did not affect the general understanding in Britain of the value of mathematics, which in the nineteenth century remained very much as John Dee and John Wallis had described in the sixteenth and seventeenth centuries – namely, as being essential for dealing with practical problems concerning multitude and magnitude, and for formulating theories concerning the physical world.

Arithmetic, geometry and algebra comprised the principal components of mathematics, and belief in the truth of mathematical results was supported by the extent to which these components, and the rules that governed them, had been engendered by nature and natural processes. However, in the eighteenth and nineteenth centuries there were introduced unintuitive and controversial mathematical concepts such as complex numbers, infinitesimals and the infinite, quaternions, other non-commutative algebras, and non-Euclidean geometries; and these demanded a re-evaluation of the grounds for believing that mathematics embodied truth. Nevertheless, such concerns did not cause popular alarm concerning the structural integrity of bridges, the financial stability of insurance companies, the accuracy of ephemerides or the efficiency of telegraph cables; and no one (as far as we know) ceased to use mathematics in day-to-day affairs because it was becoming unnatural and unintuitive.

Although the practical value of mathematics was never doubted – and we have seen in Chapter 5 the extent to which British applied mathematicians and mathematical physicists were pre-eminent in their fields – this did not help the new pure mathematicians, who found themselves in difficulties because questions concerning utility that had obvious and affirmative answers when posed of mixed mathematics had no such answers when posed of pure mathematics in isolation. These questions addressed the relevance of pure mathematicians' activities to the utilitarian preoccupations of Victorian Britain, and they gained additional traction from the gulf that was opening up between an expanded conceptualisation of pure mathematics and the principles of natural philosophy.

Not only were pure mathematicians unable to justify their activities by an appeal to the success of the multifarious applications of mathematics, or to the 'truth' that their mathematics embodied, but also the work of a pure mathematician was a closed book to most people, including many mathematicians who worked in the applied sciences and were sometimes out of sympathy with their pure colleagues. It follows that any effective engagement of pure mathematicians with the wider world had to concentrate on presentation and rhetorical devices, rather than on the content of pure mathematics itself; but the arguments that were at first deployed did not convince. Clearly, pure mathematicians could no longer be characterised as support staff for applied scientists, and their presentation of themselves as explorers in the world of mathematics failed to address the value of their explorations for society in general, although this latter trope still sometimes reappears as an off-the-shelf explanation of pure-mathematical activity. Nor could an appeal to historical examples succeed: whatever the parallels that could be drawn, the Victorians were interested in what mathematics could do for them, not what mathematics had already done for others.

The London Mathematical Society was not formally dedicated to pure mathematics, and some distinguished applied mathematicians were members. Nevertheless, pure mathematics became its principal concern, and so it might have been expected to play a central role in providing pure mathematicians with positive public spin; but it preferred instead to maintain the status quo with a strict interpretation of its Rule 1, which enabled it to avoid any activity that was not directly connected with the promulgation of 'mathematical knowledge'. Its *Proceedings* provided a valuable service by giving British mathematicians the opportunity of publishing their findings in a home-grown journal that aspired (although in vain) to the quality of the best of those on the Continent; but whatever the nature of the output (and no British pure mathematician could be mentioned in the same breath as the greatest of the Continentals) this could do nothing to allay the concerns of the critics, for pure mathematicians' lack of involvement with useful outcomes was unavoidable.

Therefore no conventional argument or activity could answer the criticism that pure mathematicians had to face from heavyweights such as William Thomson (Lord Kelvin), George Biddell Airy and Thomas Henry Huxley. To such criticism there could be no satisfactory reply until pure mathematicians realised that utility was not the only criterion by which human activity should be judged, and that the characteristics about which their critics most complained could be made acceptable and even meritorious by deploying arguments that paralleled those advanced by Walter Pater and his followers when justifying their new view of aesthetics. It was from these arguments that the ascendancy of the pure mathematicians arose, for thereafter the comparison was not between the utility of two branches of mathematics, but between the creation and appreciation of beautiful things on the one hand, and on the other the construction of necessarily flawed mathematical models of physical processes, and the solving of mundane problems.

The consequence was that pure mathematics, becoming ever more abstract, was no longer in competition with the sciences, and indeed the values that led to the sciences being so prized would have been regarded as attainting its purity. Thus the authors of 'deep theorems',[a] which possess a very high degree of abstraction, came to be the most exalted of mathematical practitioners, because they had the greatest insight into the hidden beauty of pure mathematics. In this way, pure mathematics – until recently an inseparable component of mixed mathematics – was now to be appreciated, albeit by a very small coterie, in the same manner as were the productions of poets, painters and other artists.

[a] See page 7.

It will always be the case that new uses are found for pure mathematics. Yet our study of nineteenth-century Victorian pure mathematicians has shown that such a consideration was untenable when it comes to the justification of pure mathematical activity, as Glaisher and Russell acknowledged. Rather, it was the well-timed arrival of the aesthetic tendency that gave pure mathematicians a persona that would make their discipline acceptable in the court of public and academic opinion. The pure mathematician could present himself as a being transformed: he acted neither as a servant of the sciences, nor as an explorer in a mathematical world, nor to bolster the ambitions of a vigorous and expanding Empire, for he had become a creator and contemplator of the beautiful.

SELECT BIBLIOGRAPHY

Airy, G B. 'On a Supposed Alteration in the Amount of Astronomical Aberration of Light, produced by the Passage of the Light through a Considerable Thickness of Refracting Medium', *Proceedings of the Royal Society of London*, 20 (1871–1872), 35–39.

Airy, W (ed.). *Autobiography of Sir George Biddle Airy, K.C.B., M.A., LL.D., D.C.L., F.R.S., F.R.A.S., Honorary Fellow of Trinity College, Cambridge, Astronomer Royal from 1836 to 1881* (Cambridge: Cambridge University Press, 1896).

Allen, G. *Miss Cayley's Adventures* (New York: Putnam's, 1899).

Babbage, C. *Reflections on the Decline of Science in England, and on some of its Consequences* (London: Fellowes, 1830).

Bachem, A. 'Mathematics: From the Outside Looking In', in Björn Engquist and Wilfried Schmid (eds.), *Mathematics Unlimited: 2001 and Beyond* (Berlin: Springer, 2001).

Bacon, F. *The Twoo Bookes of Francis Bacon, Of the proficience and advancement of Learning, divine and humane* (At London, Printed for Henrie Tomes, and are to be sould at his shop at Graies Inne Gate in Holborne, 1605).

Ball, R. 'The Non-Euclidean Geometry', *Hermathena*, 3, 6 (1879), 500–541.

Barlow, P. *A new Mathematical and Philosophical Dictionary; comprising an Explanation of the Terms and Principles of Pure and Mixed Mathematics ... And an Account of the Discoveries and Writings of the most celebrated Authors, both ancient and modern* (London: G & S Robinson, 1814).

Barrow, I. *The Usefulness of Mathematical Learning explained and demonstrated: being Mathematical Lectures read in the Publick Schools at the University of Cambridge, to which is prefixed, the Oratorical Preface of our Learned Author, spoke before the University on his being elected Lucasian

Professor of the Mathematics, tr. Rev. Mr John Kirkby (London: Stephen Austen, 1734).

Becher, H W. 'Radicals, Whigs and Conservatives: The Middle and Lower Classes in the Analytical Revolution at Cambridge in the Age of Aristocracy', *British Journal for the History of Science*, 28 (1995), 405–426.

Besant, W. *Fifty Years Ago* (London: Chatto & Windus, 1888).

Blakesley, J W. *Where Does the Evil Lie? Observations Addressed to the Resident Members of the Senate on the Prevalence of Private Tuition in the University of Cambridge* (London: Fellowes, 1845).

Bollobás, B (ed.). *Littlewood's Miscellany* (Cambridge: Cambridge University Press, 1986).

Bourne, A A. 'Further Reminiscences', *The Mathematical Gazette*, X (1921), 244–246.

Bright, C. *The Story of the Atlantic Cable* (London: Newnes, 1903).

Brock, M G, Curthoys, M C. *The History of the University of Oxford, Volume VII: Nineteenth-Century Oxford*, 2 (Oxford: Clarendon Press, 2000).

Carr-Saunders, A M, Wilson, P A. *The Professions* (Oxford: Clarendon Press: 1933).

Clifford, W K. 'On the Space-Theory of Matter' [Abstract], *Proceedings of the Cambridge Philosophical Society*, II (1864–76), 158.

Clifford, W K. *Seeing and Thinking* (London: Macmillan, 1879).

Clifford, W K. *The Common Sense of the Exact Sciences* (London: Kegan Paul, 1885).

Clodd, E. *Grant Allen: A Memoir* (London: Grant Richards, 1900).

Cogan, M L. 'The Problem of Defining a Profession', *Annals of the American Academy of Political and Social Science*, 297 (1955), 105–111.

Collingwood, E F. 'A Century of the London Mathematical Society', *Journal of the London Mathematical Society*, 41, 4 (1966), 577–594.

Crilly, T. 'The Cambridge Mathematical Journal and its Descendants: The Linchpin of a Research Community in the Early and Mid-Victorian Age', *Historia Mathematica*, 31 (2004), 455–497.

Crilly, T. *Arthur Cayley: Mathematician Laureate of the Victorian Age* (Baltimore, MD: Johns Hopkins, 2006).

Dauben, J W. 'Mathematicians and World War I: The International Diplomacy of G. H. Hardy and Gösta Mittag-Leffler as reflected in their Personal Correspondence', *Historia Mathematica*, 7, 3 (1980), 261–288.

Dee, J. *The Elements of Geometrie of the most aunciente Philosopher EVCLIDE of Megara. Faithfully (now first) translated into the Englishe toung, by H. Billingsley, Citizen of London ... With a very fruitfull Præface made by M. I. Dee, specifying the chiefe Mathematicall Scièces, what they are, and wherunto commodious: where, also, are disclosed certaine new Secrets*

Mathematicall and Mechanicall, vntill these our daies, greatly missed (London: John Daye, 1570).
Delany, P. 'Russell's Dismissal from Trinity: A Study in High Table Politics', *Russell: The Journal of the Bertrand Russell Archives*, ns 6, 1 (1986), 39–61.
De Morgan, A. [Anon.] 'Polytechnic School of Paris', *Quarterly Journal of Education*, 1(1) (Jan–Apr 1831), 57–74.
De Morgan, A. 'Professional Mathematics', in *Central Society of Education, Second Publication* (London: Taylor & Walton, 1838).
De Morgan, S. *Memoir of Augustus De Morgan* (London: Longmans, Green, 1882).
Descartes, R, Miller, V R, Miller, R P. *Principles of Philosophy* (Dordrecht: Reidel, 1983).
Despeaux, S E. 'Launching Mathematical Research without a Formal Mandate: The Role of University-Affiliated Journals in Britain', *Historia Mathematica*, 34 (2007), 89–106.
Despeaux, S E. '"Very Full of Symbols": Duncan F. Gregory, the Calculus of Operations, and the *Cambridge Mathematical Journal*', in Jeremy Gray and Karen Hunger Parshall (eds.), *Episodes in the History of Modern Algebra (1800–1950)* (Providence, RI: American Mathematical Society and London Mathematical Society, 2007).
Despeaux, S E. 'Fit to print? Referee Reports on Mathematics for the Nineteenth-Century Journals of the Royal Society of London', *Notes and Records of the Royal Society*, 65 (2011), 233–252.
Despeaux, S E. 'Mathematical Questions: A Convergence of Mathematical Practices in British Journals of the Eighteenth and Nineteenth Centuries', *Revue d'Histoire des Mathématiques*, 20 (2014), 5–71.
Dubbey, J M. *The Mathematical Work of Charles Babbage* (Cambridge: Cambridge University Press, 1978).
Eddington, A. 'Joseph Larmor', *Obituary Notices of Fellows of the Royal Society*, 4 (1942).
Einstein, A. (tr. & ed. Paul Arthur Schilpp), *Autobiographical Notes* (La Salle, IL: Open Court, 1979); originally published 1949.
Elliott, E B. 'Some Secondary Needs and Opportunities of English Mathematicians', *Proceedings of the London Mathematical Society*, XXX (1898), 5–23.
Enros, P C. 'The Analytical Society (1812–1813): Precursor of the Renewal of Cambridge Mathematics', *Historia Mathematica*, 10 (1983), 24–47.
Faraday, M. 'On some new Electro-Magnetical Motions, and on the Theory of Magnetism', *Quarterly Journal of Science, Literature and the Arts*, XII (1822), 74–96.
Faraday, M. 'Experimental Researches in Electricity', *Philosophical Transactions of the Royal Society of London*, 122 (1832), 125–162.

Faraday, M. 'Thoughts on Ray-vibrations' [Letter to Richard Phillips dated 15 Apr 1846], *London, Edinburgh and Dublin Philosophical Magazine and Journal of Science*, Series 3, 28, 188 (1846), 345–350.

Faraday, M. 'On lines of Magnetic Force; their definite Character; and their Distribution within a Magnet and through Space', *Philosophical Transactions of the Royal Society of London*, 142 (1852), 25–56.

Fauvel, J, Flood, R, Wilson, R (eds.). *Oxford Figures: 800 Years of the Mathematical Sciences* (Oxford: Oxford University Press, 2013).

Fiske, T S. 'Mathematical Progress in America: Presidential Address delivered before the American Mathematical Society at its Eleventh Annual Meeting December 29, 1904', *Bulletin of the American Mathematical Society*, XI, 5 (1905), 238–246.

Fitz Gerald, G F. 'The Ether and the Earth's Atmosphere', *Science*, XIII, 328 (1889), 390.

Forbes, J D. *A Review of the Progress of Mathematical and Physical Science in more Recent Times, and particularly between the Years 1775 and 1850; being one of the Dissertations prefixed to the Eighth Edition of the Encyclopaedia Britannica* (Edinburgh: A & C Black, 1858).

Forsyth, A R. 'Presidential Address to Section A of the British Association', *Report of the Sixty-seventh Meeting of the British Association for the Advancement of Science* (London: John Murray, 1898).

Forsyth, A R. 'James Whitbread Lee Glaisher', *Journal of the London Mathematical Society*, 4, 2 (1929), 101–112.

Gandy, R O. 'Bertrand Russell, as a Mathematician', *Bulletin of the London Mathematical Society*, 5 (1973), 342–348.

Garcia, P. *The Life and Work of Major Percy Alexander MacMahon* (PhD Thesis: Open University, UK, 2006).

Glaisher, J W L. 'The Mathematical Tripos', *Proceedings of the London Mathematical Society*, XVIII (1886), 4–38.

Glaisher, J W L. 'Presidential Address to Section A', in *Report of the Sixtieth Meeting of the British Association for the Advancement of Science* (London: John Murray, 1891).

Glaisher, J W L (ed.). 'Introduction', in *The Collected Mathematical Papers of Henry John Stephen Smith M.A., F.R.S., late Savilian Professor of Geometry in the University of Oxford*, I (Oxford: Clarendon Press, 1894).

Glaisher, J W L. 'Notes on the Early History of the Society', *Journal of The London Mathematical Society*, 1, 1 (1926), 51–64.

Goldstein, C, Gray, J, Ritter, J (eds.). *Mathematical Europe: History, Myth, Identity* (Paris: Éditions de la Maison des Sciences de l'homme, 1996).

Greenhill, A G. 'Collaboration in Mathematics', *Proceedings of the London Mathematical Society*, XXIV (1892), 5–16.

Gregory, D. 'Preface', *Cambridge Mathematical Journal*, I (1837), 1–2.

Griffin, N. *Russell's Idealist Apprenticeship* (Oxford: Clarendon Press, 1991).

Hardy, G H. *Some Famous Problems of the Theory of Numbers, and in particular Waring's Problem: An Inaugural Lecture delivered before the University of Oxford* (Oxford: Clarendon Press, 1920).

Hardy, G H. 'Dr. Glaisher and the *"Messenger of Mathematics"*', *Messenger of Mathematics*, LVIII (1929), 159–160.

Hardy, G H. 'Prolegomena to a Chapter on Inequalities', *Journal of the London Mathematical Society*, 4, 1 (1929), 61–78.

Hardy, G H. 'Mathematics', *The Oxford Magazine*, 48, 5 (1930), 819–821.

Hardy, G H. *A Mathematician's Apology* (Cambridge: Cambridge University Press, 1940).

Hardy, G H. *Bertrand Russell and Trinity: A College Controversy of the Last War* ('Printed for the Author' at the University Press, Cambridge, 1942).

Helmholtz, H von. 'The Axioms of Geometry', *The Academy*, I, 5 (1870), 128–131.

Helmholtz, H von. Letter to the Editor, 'The Axioms of Geometry', *The Academy*, III, 41 (1872), 52–53.

HMSO. *Report of Her Majesty's Commissioners appointed to inquire into the State, Discipline, Studies, and Revenues of the University and Colleges of Oxford: Together with the Evidence, and an Appendix* [volume 22 of the Reports from Commissioners] (London: Her Majesty's Stationery Office, 1852).

HMSO. *Report of Her Majesty's Commissioners appointed to inquire into the State, Discipline, Studies, and Revenues of the University and Colleges of Cambridge: Together with the Evidence, and an Appendix* [volume 44 of the Reports from Commissioners] (London: Her Majesty's Stationery Office, 1852).

Hodge, W D V. 'Henry Frederick Baker (1866–1956)', *Biographical Memoirs of Fellows of the Royal Society*, 2 (November, 1956).

Honey, J R de S. *Tom Brown's Universe: The Development of the Victorian Public School* (London: Millington, 1977).

Hopkins, W. *Remarks on certain proposed Regulations respecting the Studies of the University and the Period of conferring the Degree of B.A.* (Cambridge: Deighton, 1841).

Huxley, T H. 'Scientific Education: Notes of an After-Dinner Speech', *Macmillan's Magazine*, 20, 116 (1869), 178–184.

Huxley, T H. 'The Scientific Aspects of Positivism', *The Fortnightly Review*, ns 5, 30 (1869), 653–670.

Johnson, R V. *Aestheticism* (London: Methuen, 1969).

Katz, K. *The Impact of the Analytical Society on Mathematics in England in the First Half of the Nineteenth Century*, PhD thesis (1982) for New York University (Ann Arbour, MI: University Microfilms International, 1985).

Kelsall, C. *Phantasm of an University: With Prolegomena* (London: White, Cochrane, 1814).

Larmor, J. *Aether and Matter: A Development of the Dynamical Relations of the Aether to Material Systems on the basis of the Atomic Constitution of Matter, including a Discussion of the Influence of the Earth's Motion on Optical Phenomena* (Cambridge: Cambridge University Press, 1900).

Larmor, J. 'Address by the Retiring President', *Proceedings of the London Mathematical Society*, 16, 1 (1917), 1–7.

Levi, L. 'On the Progress of Learned Societies, illustrative of the Advancement of Science in the United Kingdom during the last Thirty Years', in *Report of the Thirty-eighth Meeting of the British Association for the Advancement of Science* (London: John Murray, 1869).

Lowe, V. *Alfred North Whitehead: The Man and his Work*, 1 (Baltimore, MD: Johns Hopkins, 1985).

Mahon, B. *The Man who Changed Everything: The Life of James Clerk Maxwell* (Chichester: Wiley, 2003).

Masters, B R. 'Lord Rayleigh: A Scientific Life', *Optics and Photonics News*, 20, 6 (2009), 36–41.

Maxwell, J C. 'On Physical Lines of Force, III: The Theory of Molecular Vortices', *The London, Edinburgh and Dublin Philosophical Magazine and Journal of Science*, series 4, XXIII, CLI (1862), 12–24.

Maxwell, J C. 'On Faraday's Lines of Force', *Transactions of the Cambridge Philosophical Society*, 10, 1 (1864), 27–83.

Maxwell, J C. 'A Dynamical Theory of the Electromagnetic Field', *Philosophical Transactions of the Royal Society of London*, 155 (1865), 459–512.

Maxwell, J C. *Theory of Heat* (London: Longmans Green, 1871).

Maxwell, J C. 'Letter to D P Todd dated 19 Mar 1879', *Nature*, XXI (1879–1880), 315.

Michelson, A A. 'On the Relative Motion of the Earth and the Luminiferous Ether', *The American Journal of Science*, 3, XXII, 128 (1881), 120–129.

Michelson, A A, Morley, E W. 'On the Relative Motion of the Earth and the Luminiferous Ether', *The American Journal of Science*, 3, XXXIV, 203 (1887), 333–345.

Mordell, L J. 'Hardy's "A Mathematician's Apology"', *The American Mathematical Monthly*, 77, 8 (1970), 831–836.

Morus, I R. '"A Dynamical Form of Mechanical Effect": Thomson's Thermodynamics', in R Flood, M McCartney, A Whitaker (eds.), *Kelvin: Life, Labours and Legacy* (Oxford: Oxford University Press, 2008).

Nahin, P J. *Oliver Heaviside: The Life, Work and Times of an Electrical Genius of the Victorian Age* (Baltimore, MD: Johns Hopkins, 2002).

Parshall, K H. 'The One-Hundredth Anniversary of the Death of Invariant Theory?', *Mathematical Intelligencer*, 12, 4 (1990), 10–16.

Parshall, K H. *James Joseph Sylvester: Life and Work in Letters* (Oxford: Clarendon Press, 1998).

Parshall, K H. *James Joseph Sylvester: Jewish Mathematician in a Victorian World* (Baltimore MD: Johns Hopkins, 2006).

Parshall, K H, Rowe, D E. *The Emergence of the American Mathematical Research Community, 1876–1900: J J Sylvester, Felix Klein, and E H Moore* (Providence, RI: American Mathematical Society and London Mathematical Society, 1994).

Pater, W H. *Studies in the History of the Renaissance* (London: Macmillan, 1873).

Perkin, H. *The Rise of Professional Society: England since 1880* (London: Routledge, 1989).

Playfair, J. 'Dissertation Second: Exhibiting a General View of the Progress of Mathematical and Physical Science, since the Revival of Letters in Europe', in *Supplement to the Fourth, Fifth, and Sixth Editions of the Encyclopaedia Britannica* (Edinburgh: Archibald Constable, 1824).

Poole, J E. 'Dr J W L Glaisher: The Making of a Great Collection', in Michael Archer (ed.), *Delftware in the Fitzwilliam Museum* (London: Philip Wilson, 2013).

Reader, W J. *Professional Men: The Rise of the Professional Classes in Nineteenth-Century England* (London: Weidenfeld & Nicolson, 1966).

Rice, A, Wilson, R J. 'The Rise of British Analysis in the Early 20th Century: The Role of G H Hardy and the London Mathematical Society', *Historia Mathematica*, 30, 2 (2003), 173–194.

Rice, A C, Wilson, R J, Gardner, J H. 'From Student Club to National Society: The Founding of the London Mathematical Society in 1865', *Historia Mathematica*, 22, 4 (1995), 402–421.

Richards, J. *Mathematical Visions: The Pursuit of Geometry in Victorian England* (San Diego, CA: Academic Press, 1988).

Riemann, B. (tr. W K Clifford), 'On the Hypotheses which lie at the Bases of Geometry: Plan of the Investigation', *Nature*, 8, 183 (1873), 14–17.

Riemann, B. (tr. W K Clifford), 'On the Hypotheses which lie at the Bases of Geometry: Application to Space', *Nature*, 8, 184 (1873), 36–37.

Roberts, S. 'Remarks on Mathematical Terminology, and the Philosophic Bearing of recent Mathematical Speculations concerning the Reality of Space', *Proceedings of the London Mathematical Society*, XIV (1882), 5–15.

Rose, H J. 'Introduction to the Pure Sciences', in *Encyclopaedia Metropolitana, or, Universal Dictionary of Knowledge, on an Original Plan: comprising the Twofold Advantage of a Philosophical and an Alphabetical Arrangement with appropriate Engravings*, I (Pure Sciences I) (London: Fellowes, 1845).

Rota, G C. 'Introduction', in George E Andrews (ed.), *Percy Alexander MacMahon: Collected Papers*, I (Cambridge MA: MIT Press, 1978).

Roth, L. 'Old Cambridge Days', *American Mathematical Monthly*, 78 (1971), 223–236.

Rothblatt, S. *The Revolution of the Dons: Cambridge and Society in Victorian England* (London: Faber, 1968).

Russell, B. *An Essay on the Foundations of Geometry* (Cambridge: Cambridge University Press, 1897).

Russell, B. *The Principles of Mathematics* (Cambridge: Cambridge University Press, 1903).

Russell, B. 'The Study of Mathematics', *The New Quarterly*, 1, 1 (1907), 31–44.

Russell, B. 'Mathematics and the Metaphysicians', in *Mysticism and Logic and Other Essays* (London: Longmans, 1918).

Russell, B. 'Why I took to Philosophy', in *Portraits from Memory, and Other Essays* (London: Allen & Unwin, 1956).

Russell, B. *The Autobiography of Bertrand Russell: 1872–1914*, I (London: Allen & Unwin, 1967).

Russell, C. *Science and Social Change, 1700–1900* (London: Macmillan, 1983).

Simpson, R. *How the PhD came to Britain: A Century of Struggle for Postgraduate Education* (Guildford: The Society for Research into Higher Education, 1983).

Small, I. *Conditions for Criticism: Authority, Knowledge, and Literature in the Late Nineteenth Century* (Oxford: Clarendon Press, 1991).

Smith, H J S. 'On the Present State and Prospects of some Branches of Pure Mathematics', *Proceedings of the London Mathematical Society*, VIII (1876), 6–29.

Smith, J, Stray, C (eds.). *Teaching and Learning in Nineteenth-Century Cambridge* (Woodbridge: Boydell Press, 2001).

Snow, C P. 'Foreword', in G H Hardy, *A Mathematician's Apology* (Cambridge: Cambridge University Press, 1967).

Spottiswoode, W. 'Remarks on some Recent Generalizations of Algebra', *Proceedings of the London Mathematical Society*, IV (1871), 147–164.

Spottiswoode, W. 'Address', in *Report of the Forty-eighth Meeting of the British Association for the Advancement of Science* (London: John Murray, 1879).

Stokes, G G. 'On the Constitution of the Luminiferous Aether', *The London, Edinburgh, and Dublin Philosophical Magazine and Journal of Science*, XXXII (1848), 343–349.

Stray, C. *Classics Transformed: Schools, Universities, and Society in England, 1830–1960* (Oxford: Clarendon Press, 1998).

Sylvester, J J. 'Presidential Address to Section "A" of the British Association', in *Report of the Thirty-eighth Meeting of the British Association for the Advancement of Science* (London: John Murray, 1869).

Sylvester, J J. 'Address', *Report of the Thirty-ninth Meeting of the British Association for the Advancement of Science* (London: John Murray, 1870).

Tait, P G. 'James Clerk Maxwell', *Proceedings of the Royal Society of Edinburgh*, 10 (1878–1880), 331–339.

Tanner, J R (ed.). *The Historical Register of the University of Cambridge, being a Supplement to the Calendar with a Record of University Offices, Honours and Distinctions to the Year 1910* (Cambridge: Cambridge University Press, 1917).

Thompson, S P. *The Life of William Thomson Baron Kelvin of Largs*, I (London: Macmillan, 1910).

Thomson, H B. *The Choice of a Profession. A Concise Account and Comparative Review of the English Professions* (London: Chapman & Hall, 1857).

Thomson, J J. 'Dr. Glaisher', *The Cambridge Review*, L, 1228 (1929), 212–231.

Thomson, W. 'On the Age of the Sun's Heat', *Macmillan's Magazine*, V (1861–1862), 388–393.

Thomson, W. 'On the Secular Cooling of the Earth', *Transactions of the Royal Society of Edinburgh*, XXIII (1862), 157–169.

Thomson, W. 'On Geological Time', in Sir William Thomson (Baron Kelvin), *Popular Lectures and Addresses*, 2 (London: Macmillan, 1894).

Turner, H H. 'James Whitbread Lee Glaisher', *Monthly Notices of the Royal Astronomical Society*, LXXXIX, 4 (1929), 300–308.

University of Cambridge. *The Student's Guide to the University of Cambridge* (Cambridge: Deighton, Bell, 1863).

Venn, J A (comp.), *Alumni Cantabrigienses: A Biographical List of all known Students, Graduates, and Holders of Office at the University of Cambridge from the Earliest Times to 1900*, II, (Cambridge: Cambridge University Press, 1940–1954).

Wallis, J. 'Dr. Wallis's Account of some Passages of his own Life', in Thomas Hearne, *Peter Langtoft's Chronicle, (as illustrated and improv'd by Robert of Brunne) from the Death of Cadwaladar to the End of K. Edward the First's Reign. To which are added ... other curious Papers ...* (Oxford: Printed at the Theater, 1725).

Walter, T. *A new Mathematical Dictionary, containing the Explication of all the Terms in Pure and Mixed Mathematics ... To which is prefixed, the Elements of Geometry* (London: The Author, 1761).

Warwick, A. 'The Laboratory of Theory, or, What's Exact about the Exact Sciences?', in M Norton Wise (ed.), *The Values of Precision* (Princeton, NJ: Princeton University Press, 1995).

Warwick, A. *Masters of Theory: Cambridge and the Rise of Mathematical Physics* (Chicago: University of Chicago Press, 2003).

Whewell, W. 'Review of "On the Connexion of the Physical Sciences. By Mrs Somerville"', *The Quarterly Review*, 51 (1834), 54–68.

Whewell, W. *Thoughts on the Study of Mathematics, as part of a Liberal Education*, Second Edition (Cambridge: Deighton, 1836).
Whewell, W. *On the Principles of an English University Education* (London: Parker, 1837).
Whittaker, E T. *A History of the Theories of Aether and Electricity from the Age of Descartes to the Close of the Nineteenth Century* (London: Longmans, Green, 1910).
Whittaker, E T. 'Andrew Russell Forsyth', *Obituary Notices of Fellows of the Royal Society*, 4 (November, 1942).
Wiener, N. *Ex-Prodigy: My Childhood and Youth* (New York: MIT Press, 1964).
Wilson, J M. 'The Early History of the Association: Or, the Passing of Euclid from our Schools and Universities, and how it came about. A Story of Fifty Years ago', *The Mathematical Gazette*, X (1921), 239–244.
Worsley, T. *Christian Drift of Cambridge Work: Eight Lectures recently delivered in Chapel on the Christian Bearings of Classics Mathematics Medicine and Law Studies prescribed in its Charter to Downing College* (London: Macmillan, 1865).
Young, J R. *An Introductory Lecture, delivered at the opening of the Mathematical Classes of Belfast College, November 12, 1833* (London: J Souter, 1833).
Young, T. 'Outlines of Experiments and Inquiries Respecting Sound and Light', *Philosophical Transactions of the Royal Society of London*, 90 (1800), 106–150.
Young, T. 'The Bakerian Lecture. Experiments and Calculations relative to Physical Optics', *Philosophical Transactions of the Royal Society of London*, 94 (1804), 1–16.
Young, T. Letter to François Arago dated 12 Jan 1817, in George Peacock (ed.), *Miscellaneous Works of the late Thomas Young*, 1 (London: John Murray, 1855).

INDEX

Abbott, Edwin Abbott, 166
aberration, stellar, 150, 151, 183
academic, *defined*, 56
Academy of Sciences (France), 17, 131–132
action at a distance, 148
Adams, Walter Marsham, 94
Addison, Joseph, 192
Advancement of Learning (Bacon), 12
aesthetes and aesthetics, 214–218
aestheticism, *defined*, 216
aether, 149, 159–164, 170, 183
    elasticity of, 151–152, 160, 164, 172
aether-drag, 161–163
Airy Transit Circle, 181
Airy, George Biddell, 18, 35, 180–184, 196, 221
    pure mathematicians, excoriates, 202–203
Albert Medal, 196
Albert, Prince of Saxe-Coburg and Gotha, 36
algebra, 25, 28–30, *See also* invariants; matrices; quadratic equations; quaternions
Allen, Grant, 229
American Mathematical Society, 57, 88
Analytical Society, 18, 22

analytics, 19, 29–32
applied mathematician *(word)*, origin of, 6
applied mathematics, 2, 70, 235
Aristotle, *See* scholasticism
Asiatic Society, 66
Association for the Improvement of Geometrical Teaching, 80–82
associativity, *defined*, 38
authors, mathematical, 54–56

Babbage, Charles, 18, 22, 24
Bachem, Achim, 7
Bacon, Francis (Lord Verulam), 12
Baker, Henry Frederick, 101, 127–129
Ball, Robert, 165
Ballade of the Girton Girl (Lang), 229
Barlow, Peter, 24
Barrow, Isaac, 12, 14
Basset, Alfred, 99
Beeching, Henry Charles, 131
Besant, Walter, 192, 195
Besant, William Henry, 123, 226
Birmingham, University of, 91
Boltzmann, Ludwig, 86
Bólyai, János, 165
Bompiani, Enrico, 198
Boole, George, 39–40, 231

Boolean algebra, 39
Bouch, Thomas, 184
Boyle, Robert, 191
Bradley, James, 150, 183
Bristol, University of, 91
British Association for the Advancement of Science, 4, 6, 50, 68, 197
  Exeter meeting, address by Sylvester, 203–207
  Southport meeting
    controversy concerning, 218
    Presidential Address by Cayley, 224–226
Bronowski, Jacob, 129
Budget of Paradoxes (De Morgan), 52
Burlington House, 66

calculus, 16, 17, 40
  differential coefficient, 20–21
  discovery of, dispute concerning, 3, 16, 26
  notation, 18
Cambridge [and Dublin] Mathematical Journal, 33, 35, 56, See also *Quarterly Journal of [Pure and Applied] Mathematics*
Cambridge Philosophical Society, 56, 65
Cambridge, University of, 4, 6, 55–56, 127, 235, See also Classical Tripos; Mathematical Tripos
  calculus, establishment of, 16–18
  careers after graduating, lack of, 21, 35, 43, 74
  competition between students, encouraged, 15, 43, 103, 209
  complex analysis, ignored, 117
  liberal education, 21, 31–32
  London Mathemtical Society, dominated by graduates of, 72
  Lowndean chair, 23
  Lucasian chair, 12, 22–23, 172
  mathematical tradition, 44, 140, 145
  Plumian chair, 23
  Royal Commission for, 28, 35, 209
  Sadleirian chair, 39, 113

senior lectureship in mathematics, instituted, 115
Cantor, Georg, 102
Cayley, Arthur, 113, 120, 121, 145, 206
  Airy, dispute with, concerning pure mathematics, 202
  BAAS Southport meeting
    President of, 218
    Presidential Address, 224–226
  De Morgan medal, awarded first, 85–86
  intuition, use of, 126, 197
  LMS, member of, 66, 83
  matrices, theory of, developed by, 9, 38–39
  productivity, phenomenal, 75
Chasles, Michel, 68–69, 98
Chemical Society, 66
Clarendon Commission, 15
classical mechanics, *defined*, 150
Classical Tripos, 21, 41
classics, 15, 41–44
Clausius, Rudolf, 155
Clifford, William Kingdon, 83, 160, 165, 166
  non-Euclidean geometry, speculations concerning, 170–172
Clifton, Robert Bellamy, 63
Collingwood, Sir Edward, 60
Common Sense of the Exact Sciences (Clifford), 166
commutativity, *defined*, 38
complex analysis, 117
  defined, 111
Comptes Rendus Hebdomadaires des Séances de l'Académie des Sciences, 131
conic sections, 64
constructive proof, 127
Cooper's Hill Engineering College, 136
corpuscular theory of matter, See Descartes, René
couple, *defined*, 164
Cousin, Victor, 215
Coxeter, Harold Scott MacDonald, 129

Cramér, Harald, 237
Crelle, August, 119
Crelle's Journal, See under *Journal für die reine und angewandte Mathematik*
Crookes, William, 180
Cunningham, Allan, 92

d'Alembert, Jean de la Ronde, 17
Darwin, Charles, 177, 229
Darwin, Sir George, 98, 234–236
Davy, Sir Humphrey, 153
De Morgan medal, 60, 85–87, 102
De Morgan, Augustus, 15, 27, 52–53, 80, See also De Morgan's laws
   calculus, remarks concerning, 17, 21
   career, 40–41
   LMS, Presidential Address to, 64–66, 68, 101–102
   professional mathematicians, defines, 32
De Morgan, George Campbell, 57, 68
De Morgan, Sophia, 61, 68
De Morgan's laws, 40
Dedekind, Richard, 102
Dee, John, 11
Department of Artillery Studies (RMA), 135
derivative, See under calculus *sub* differential coefficient
Descartes, René, 147–148
Deutsche Mathematiker-Vereinigung, 57
differential coefficient, See under calculus
diffraction, 148
double refraction, 150
Du Chaillu, Paul, 206
du Maurier, George, 218
Duhem, Pierre, 20
duplication of a cube, 68
Durham, University of, 21

Eddington, Sir Arthur, 163
Edinburgh, University of, 74
Educational Times, 80

Einstein, Albert, 9, 157, 159
Eisenstein, Gotthold, 131, 206
elasticity, See under aether
electromagnetic field, *defined*, 157
electromagnetism, 146, 152–154, See also under Maxwell, James Clerk
Elements of Geometry (Legendre), 168
Elliott, Edwin Bailey, 76, 88, 101
ether, See aether
Eton College, 16
Euclid's *Elements*, 2, 63, 80, 166–167
Euler, Leonhard, 17

Faraday, Michael, 152–154
   electromagnetism, advances in electromagnetic induction, 153–154
   lines of force, 154
   rotatory motion, 153
   religious beliefs, 154
Fawcett, Philippa, 15
Ferrers, Norman, 120
fifth postulate (Euclid), See parallel postulate
finite series, *defined*, 20
Fiske, Thomas, 57, 87
FitzGerald, George, 162
Fitzwilliam Museum, Cambridge, 112
Fizeau, Hippolyte, 159, 160
Flatland; A Romance of Many Dimensions (Abbott), 165
fluxions and fluents, 3, 16, 17, 18
Forbes, James David, 201
Forsyth, Andrew Russell, 54
   BAAS, Toronto meeting of, Presidential Address, 227–229
   Glaisher, opinions concerning, 112–114
   intuition, disastrous consequence of relying on, 198
   Sadleirian chair, secures, 116, 117
Foucault, Léon, 160
Fourier, Jean-Baptiste Joseph, 5
Franklin, Christine Ladd, 98
Frege, Gottlob, 102

Frend, William, 30
Fresnel, Augustin-Jean, 151

Gardiner, Martin, 71
Garstang, Thomas James, 97
gases, kinetic theory of, 155–157
Gaskin, Rev. Thomas, 28
Gauss, Carl Friedrich, 122, 165
general theory of relativity, 9, 147, 165
gentleman, origin and defining features of the, 191–192
geometry, *See also* Euclid's *Elements*
  elementary, Board of Education's circular 711 on, 97
  non-Euclidean, 165–171
George III, King, 192
Germany
  attractiveness of, for postgraduates, 87–88
  pure mathematics, superior to Britain in, 132
Gibbs, Josiah Willard, 163
Gilbert and Sullivan, 217–218
Girton Girls, mathematics, supposed fondness for, 229
Glaisher, James *(father of JWL)*, 112, 121
Glaisher, James Whitbread Lee, 54, 70, 72, 77, 85, 111–127
  addresses
    BAAS, Leeds meeting, presidential, 122–125
    LMS, presidential, 121–122
  Astronomer Royal, offered post of, 115
  books, unable to publish, 116, 118
  calculation, love of, 112
  committees and councils, service on, 120–121
  elliptic functions, delight in, 111, 114
  journals
    contributions to, 118
    Messenger of Mathematics, 112, 118–120
    Quarterly Journal of [Pure and Applied] Mathematics, 118–120

  subsidises out of own pocket, 118
  pottery collection, 112, 116
  pure mathematics, conflicted sentiments concerning, 123–124, 127
  pure sciences, school of, regrets non-existent at Cambridge, 123
  research
    commitment to, 112
    Continental practice, preferred, 121
    outdated methods of, 113, 133
    Sadleirian chair, fails to secure, 115
    teacher, success as, 114–115
Glasgow, University of, 74, 173
Goodwin, Harvey, 35
Gordan, Paul, 137, 198
Grand Prix des Sciences Mathématiques, 131
Great Exhibiton of 1851, 5
Great Transit Circle, *See* Airy Transit Circle
Greenhill, Alfred George, 77, 125, 126, 136
Gregory, Duncan F, 33
Grossmann, Marcel, 9
Grossmith, George, 217
Grote, George, 43
groups, theory of, 138

Halmos, Paul, 125
Hamilton, William Rowan, 38
Hardcastle, Frances, 98
Hardy, Godfrey Harold, 99–100, 197, 236–241
  abstracts, urges introduction of, 95
  Germany, valued relations with, 99–100
  Glaisher, opinions concerning, 113, 120
  Glaisher's journals, criticised, 118
  Littlewood, partnership with, 99, 100, 138, 198, 236–237
  Oxford University, move to, 237–239
  pure mathematics
    logic now subsumed under, 239
    promoter of, 99

reality of, 208
rigour, concern with, 198, 236
uselessness of, causes depression, 6, 240
Savilian professor, inaugural lecture as, 6, 239–240
Harrow School, 16
Heaviside layer, 163
Heaviside, Oliver, 163
Helmholtz, Hermann von, 165, 202
Henrici, Olaus, 77, 88
Hermite, Charles, 131
Herschel, John, 18, 22, 24
Hertz, Heinrich, 159
Hilbert, David, 86, 91, 125, 137
Hirst, Thomas Archer, 88
  AIGT, involvement with, 81
  Journal of, 59
  LMS, involvement with, 58, 61–63, 67–68, 83
Hobson, Ernest William, 88, 117
Hooke, Robert, 148–149, 151
Hopkins, William, 33
Hutton, Charles, 25, 82
Huxley, Thomas Henry, 203, 206, 229
Huygens, Christian, 149

infinite series, 29–30
  defined, 20
infinitesimals, 16, 19
Ingham, Albert, 237
Institution of Mechanical Engineers, 71
interferometry, 161
International Catalogue of Scientific Periodicals, 96
International Meridian Conference (1884), 181
International Research Council, 100
International Union of Mathematical Societies, 99
Intrinsic Geometry of Ideal Space (Forsyth), 198
intuition, 31–32, 126, 166, 197–198
invariants, theory of, 83, 103, 137

Investigation of the Laws of Thought (Boole), 39
Italy, abolution of Euclid in, 80
Ivory, James, 22

Jacob, Sydney Montague, 93
Jacobi, Carl, 5, 126
Jeans, James, 115
Jellaby Postlethwaite (*fictional*), 218
jelly, 160–161
Johnson's *Dictionary*, 13
Jordan, Camille, 131
Joule, James Prescott, 174
Journal de Mathématiques Pures et Appliquées, 22, 119
Journal für die reine und angewandte Mathematik, 22, 119
Journal of the London Mathematical Society, 100
journals, 22, 52, See also *individual titles*
Jowett, Benjamin, 130
junior optimes, *defined*, 15

Kant, Immanuel, 24, 168
Kelsall, Charles, 200
Kelvin, Lord, *See* Thomson, William
King's College (London), 21, 26–27
Kirkman's schoolgirl problem, 52
Klein, Felix, 57, 86, 87, 119
  seminar system, 125
Klinkerfues, Ernst, 183
Knight, Thomas, 22
Kohlrausch, Friedrich, 159

Lacroix, Silvestre François, 18, 20
Lady's and Gentleman's Diary, 52
Landau, Edmund, 237
Lang, Andrew, 229
Lang, Serge, 7
Laplace, Pierre-Simon, 17
Larmor, Joseph, 88, 90
  Baker, confidant of, 128
  elasticity, explanations of, 161, 172
  MacMahon, describes career of, 136, 137

Laws of Verse (Sylvester), 203
Lax, William, 23
Leeds, University of, 91
Legendre, Adrien-Marie, 168
Leibniz, Gottfried, 3, 17, 19, 191
Levi, Leone, 50
liberal education, *See under* Cambridge, University of
light, *See also under* Maxwell, James Clerk
   Descartes' corpuscular theory, 148
   particle theory, 149, 160
   speed of, 149, 158–159
   wave theory, 148–152, 183
limit of a series, 30
   defined, 20
Lindemann, Ferdinand von, 67
Liouville, Joseph, 119
Liouville's journal, *See Journal de Mathématiques Pures et Appliquées*
Littlewood, John Edensor, 119, 240
   Hardy, partnership with, 99, 100, 138, 198, 236–237
   non-existence of, purported, 237
Liverpool, University of, 91
Lobachevski, Nikolai, 165
Lodge, Oliver, 180
logic, 8, 31, 39
Lois Cayley (*fictional*), 229
London Mathematical Society, *See also Proceedings of the London Mathematical Society*
   applied mathematics, attitude concerning, 69–70
   badge, 61–63, 102
   Council, 92
   finances, 80, 96–97
   Germany, relations with, 99–100
   incorporation, 89
   international congresses, involvement in, 87–89, 98
   library, 66
   meetings
      first, 57–60, 64–66
   social function of, 76–77, 83, 101, 140
   membership, 71–75, 90, 93–94, 97–98
      honorary, 68, 83, 84, 90
      life, 91
      overseas, 71
      women, 98
      wranglers, 73–74
   naming of, 60, 67
   professional body, is not a, 194–196
   refereeing of papers, 67
      by a woman, 98
   role, 81–82, 95–97
   royal charter, 61, 75, 89
London University, 21, 26–27, 40, *See also* London, University of
London, University of, 26–27, 55–56, 74, *See also* King's College (London); University College (London)
longitudinal waves, *defined*, 150
Lorentz, Hendrik, 162
Love, Augustus Edward Hough, 96
Lowndean chair (Cambridge), *See under* Cambridge, University of
Lubbock, Sir John, 66
Lucas, Vrain-Denis, 68–69
Lucasian chair (Cambridge), *See under* Cambridge, University of
Luttrell, Narcissus, 13
Lytton, Henry, 217

MacCullagh, James, 152, 160, 164
Macfarlane, Alexander, 209
Mackay, John, 88
MacMahon, Percy, 85, 92, 126, 134–138
Macmillan & Co, 54, 198
Manchester New College, 57
Manchester, University of, 91
Mathematical Analysis of Logic (Boole), 39
Mathematical Association, *See* Association for the Improvement of Geometrical Teaching

Mathematical Gazette, 98
Mathematical Tripos, 36, 55, 77, 184
  competitive ethos, 103, 209
  pre-eminent examination at Cambridge, 15, 41
  reform of, urged by Glaisher, 121, 127
Mathematician, 64
Mathematician's Apology (Hardy), 240
mathematicians, British, 32, 103
  barriers of nationality and language, affected by, 138–139
  ignorant of Continental work, 3, 103, 138
  role of, 5, 26
mathematics, See also applied mathematics; mixed mathematics; practical mathematics; pure mathematics; speculative mathematics
  applications of, explained by De Morgan, 64–65
  Indian Civil Service, not valued in, 93
  practice, defined as a, 8
  scope of, 5, 11–15, 185
Mathematische Annalen, 133
matrices, 9, 38–39
Maxwell, James Clerk, 90, 154–159, 161, 163, 172
  aether, accepted existence of, 157
  dielectrics, 157
  displacement current, 158
  electromagnetic waves
    existence of, inferred, 158
    light is a manifestation of, 159
    speed of propagation, calculated, 159
  thermodynamics, work on, 155–157
Maxwell's demon, 156
Maxwell's equations, 157–159, 163
Mechanism of the Heavens (Somerville), 17
Merrifield, Charles Watkins, 83, 195
Messenger of Mathematics, 56, 115, 118–120
metaphysics, 19, 150

Michelson, Albert, 161
Michelson–Morley experiment, 161–163
Minkowski, Herman, 131
Miss Cayley's Adventures (Allen), 229
Mittag-Leffler, Magnus Gösta, 86
mixed mathematics, 2, 145
  scope of, 11–15
models
  imaginary mechanical, 150, 160, 163–164
  mathematical (*equations*), 150, 159, 172
  mathematical (*physical objects*), 77, 87
Morland, Samuel, 13
Morley, Edward, 162
Mrs Warren's Profession (Shaw), 210
Munich Exhibition of Mathematical Models, Apparatus and Instruments, 88

natural philosopher (*word*), 4
natural philosophy, 3, 4, 145
Nature, 98, 220
Neale, Cornelius, 17
negative quantities, rejection of, 31
Neptune (*planet*), 183, 185
New Mathematical Dictionary (Walter), 14
New Quarterly, 231
New York Mathematical Society, See American Mathematical Society
Newcomb, Simon, 183
Newton, Isaac, 2, 148–149
Newton's rings, 148
Newtonian mechanics, 166, 169
non-constructive proof, 127
non-Euclidean geometry, See under geometry
numbers, theory of, 9, 115, 131, 239

Oldenburg, Henry, 191
Ørsted, Hans Christian, 153
Oxford Mathematical Society, 140

Oxford, Cambridge, and Dublin Messenger of Mathematics, See Messenger of Mathematics
Oxford, University of, 15, 21, 41
  BAAS Southport meeting, controversy concerning, 219
  LMS, graduates as members of, 74
  mathematics, relationship with, 34, 209
  Royal Commission for, 34
  Savilian chair, 11, 74, 139
    Hardy's move to, 237–239

parallel postulate, 166–169
Pascal, Blaise, 20
Pascal's theorem, 64
Pater, Walter Horatio, 214–215, 216, 221, 222
Patience, or Bunthorne's Bride (Gilbert and Sullivan), 217–218
Peacock, George, 18, 23, 25
  Principle of the Permanence of Equivalent Forms, 37–38
Peano, Giuseppe, 102
Pearson, Karl, 91, 170
Pedoe, Daniel, 129
Peel, Sir Robert, 36
peers, sons of, special treatment at Cambridge, 21
Pendlebury, Richard, 55
Perigal, Henry, 76
permutations and combinations, 52
Philosophical Transactions (Royal Society), 22
pi, record-breaking calculation of, 103
Platonism, mathematical, 206
Playfair, John, 26
Playfair's axiom, 168
Plenum, See Descartes, René
Plumian chair (Cambridge), *See under* Cambridge, University of
Poe, Edgar Allan, 204
Poincaré, Henri, 86, 91
polarisation, 151
poll men, *defined*, 15

Pollock, Sir Frederick, 17
Polytechnic School of Paris, 32
Popular Astronomy (Airy), 183
practical mathematics, 14
Practical Suggestions on Mathematical Notation and Printing (LMS), 95
Principia Mathematica (Whitehead and Russell), 40
Principles of Natural Philosophy (Descartes), 147
Proceedings of the London Mathematical Society, 56, 78, 83, 90, 92
  abstracts, introduction of, 95–96
  Cambridge graduates, principal contributors, 76
  papers
    content, 73, 78, 90
    length, 76
    refereeing, 67, 98
    printing, cost of, 96
    typesetting, 94–95
professionalisation, 50, **190–199**, 216
  of pharmacy, 194
  of teaching, 195
professions, history of, 191–193
projective geometry, 169
proof
  description of, by Whitehead and Russell, 199
  methods of, 127
psychology, Victorian, needs mechanical models, 150
Punch, or the London Charivari, 218
pure mathematicians
  British, waste time on calculations, 126
  characterised as
    artists, 221
    explorers and natural historians, 206–208
    superior beings, 227, 235
  criticisms of, 201–203
  genesis of, 3, 6, 145
  role of, 51, 146, 185

pure mathematics
   art, characterised as, 220–224
   classics, compared with the, 7, 42–44
   dangers of, 200
   research in, easier at Oxford, 130
   scope of, 2–3, 8, 12–15, 25, 123–124
   superfluity of, 200
   symbolic logic, grounded in, 231
Pure Mathematics (Hardy), 236
Purloined Letter (Poe), 204

quadratic equations, 2, 28
quadrivium, 11
Quarterly Journal of [Pure and Applied] Mathematics, 56, 64, 118–120, 132, See also *Cambridge [and Dublin] Mathematical Journal*
quaternions, 38
Queen's College (Galway), 74

Ramanujan, Srinivasa, 137, 236
Ranyard, Arthur Cowper, 57, 60, 62, 72
Rayleigh, Lord, *See* Strutt, John William
real analysis, *defined*, 111
real numbers, *defined*, 38
reasoning, styles of, 19–20, 25, 32, 54, 204–206
relativity, *See* special theory of relativity; general theory of relativity
Remarks on a supposed Error in the Elements of Euclid (Lax), 23
Riemann, Bernhard, 117, 133, 165, 169
Roberts, Samuel, 66, 77, 83
   controversy, wishes to avoid, 82, 83
   De Morgan medal
      awarded, 86–87
      criteria for award of, outlines, 85
Römer, Ole, 149
Rose, Henry John, 201
Roth, Leonard, 113
Routh, Edward John, 63, 90, 112, 178
Royal Academy of Arts, 192
Royal Astronomical Society, 53, 66, 73
Royal Military Academy, 25, 64, 82

MacMahon, career at, 85, 134
Sylvester, professor at, 66, 135, 203
Royal Society, 57, 132, 154, 173
   rich dilettantes, populated by, 24
Ruskin, John, 197, 217
Russell, Bertrand, 38, 39, 102, 236
   Kant, attempts to restore epistemology of, 168
   pure mathematics
      beauty of, discourses upon, 230–234
      describes, 8

Salisbury, Lord, 219
Salmon, George, 220–224, 232
Sandeman, Archibald, 30
Sandemanians, 154
Savilian chair (Oxford), *See under* Oxford, University of
scholasticism, 147
schools, failure to teach mathematics in, 15, 55
Schubert, Hermann Cäsar Hannibal, 85
scientific societies, 50–51
scientist (*word*), 4, 145
scientists, genesis of, 5, 169
Scott, Charlotte Angas, 15, 98
Scottish universities, 21
Seeley, John Robert, 41, 43, 128
seminars, 103, 121, 128
   Klein, practice of, concerning, 125, 140
senior optimes, *defined*, 15
senior wrangler, *defined*, 15
seven liberal arts, 11
Shanks, William, 103, 232
Sharpey, William, 66
Shaw, George Bernard, 210
Shaw, Peter, 14
Sheepshanks, Richard, 22, 43
Sheffield, University of, 91
Shilling Manual of Trigonometry (Adams), 94
Smith, Henry John Stephen, 74, 76, 83, 121, **129–134**
   anti-utilitarian views, 208–209

Smith, Henry John Stephen (cont.)
  Continental appreciation of, 130, 132
  Grand Prix of 1883, embarrassing confusion over, 131–132
Smith, James, 68
Smith, Robert, 17
Smith's Prizes, 17
Société Mathématique de France, 57, 78
societies
  mathematical, 52–53, 56–57, *See also* London Mathematical Society
  professional, 90, 194
Somerville, Mary, 17
special theory of relativity, 9, 147, 183
speculative mathematics, 14
Spitalfields Mathematical Society, 52
Spottiswoode, William, 37, 38
squaring the circle, 67
St John's College (Cambridge), 74
Stanhope, Earl of, 17
Stokes, George Gabriel, 121, 161, 172–173
Stokes's theorem, 173
Story, William, 87
Strand Magazine, 229
Strutt, John William (Lord Rayleigh), 70, 90, 178–180
  London Mathematical Society, gift to, 179
  psychical research, interest in, 180
Student's Guide to the University of Cambridge (Seeley), 41
Studies in the History of the Renaissance (Pater), 214–215, 222
Suggestions for Notation and Printing (LMS), 95
Sur le Calcul Différentiel et Intégral (Lacroix), 18
Sussex, Duke of, 24
Sylvester, James Joseph, 76, 81, 165, 169
  Cambridge degree, could not be awarded, 204
  Cayley, assistance from, declined, 126

LMS
  De Morgan medal, awarded second, 85–86
  early member of, 66
  president of, 66
  pure mathematics, defends practice of, 203–207
  Royal Military Academy, professor at, 135
  Savilian chair, appointed to, 139
  symbols, enthusiasm for, 27–28, 29–30

Tait, Peter Guthrie, 126, 157, 176
Taylor expansion, 20
teaching, professionalisation of, *See under* professionalisation
tensor calculus, 9
terminology, changes in, 4, 24–25, 71, 145–146
Theory of Functions of a Complex Variable (Forsyth), 113, 117, 139
Theory of Heat (Maxwell), 156
Theory of Sound (Rayleigh), 179
thermodynamics, second law of, 155–157
Thomson, Henry Byerley, 192
Thomson, Joseph John, 115, 178, 238
Thomson, William (Lord Kelvin), 90, 121, 173–178, 196
  aether, mechanical model of, 163–164
  Atlantic Telegraph Company, involvement with, 174–175
  Earth, age of the, 176–177, 235
  energy, develops concept of, 174
  mariners' compass, improvements to, 176
  mirror galvanometer, improvements to, 175
  pure mathematicians, excoriates, 202
  Stokes's theorem, responsible for, 173
Thoughts on the Study of Mathematics as part of a Liberal Education (Whewell), 31
Todd, David Peck, 161
Todhunter, Isaac, 54, 132

Toplis, James, 17
Traité de Méchanique Céleste (Laplace), 17
Traité du calcul différentiel et du calcul intégral (Lacroix), 20
transverse waves, *defined*, 150
Treatise on Differential Equations (Forsyth), 54
Treatise on Electricity and Magnetism (Maxwell), 157
Treatise on Natural Philosophy (Thomson and Tait), 176
trigonometry, analytical, 117
Trinity College (Cambridge), 74, 114
Trinity College (Dublin), 35, 74, 223
Tripos, *See* Classical Tripos; Mathematical Tripos
trisection of an angle, 68
trivium, 11
truth, mathematical, 8, 38, 82, 140
  natural processes as validators of, 30, 51, 199, 243
  Whewell, opinions of, concerning, 32, 36
Tucker, Robert, 59, 61, 81
Turton, Thomas, 23

universities, 55–56, 91–92, 146, *See also* individual universities
University College (London), 21, 26–27, 40
  LMS, involvement with, 57–59, 61, 66
University College Mathematical Society, *See* London Mathematical Society
University College School, 58

University Hall (London), 57
Ussher, James, 176

vectors, 154, 163
Victoria, University of, 91
vortices, 148, 164
Vrain-Lucas, Denis, *See* Lucas, Vrain-Denis

Wales, University of, 91
Walker, John James, 83
Wallis, John, 11
Walter, Thomas, 13–14, 19, 25
Waring, Edward, 17
Waring's problem, 239
Weber, Wilhelm, 159
Weierstrass, Karl, 117, 133
Whewell, William, 31–32, 35–36
Whistler, James McNeill, 215, 217
Whitehead, Alfred North, 40, 102, 115
Whitehouse, Edward Orange Wildman, 174–175
Whittaker, Edmund, 117, 164
Wiener, Norbert, 241
Wilde, Oscar, 215, 217
Wilson, James Maurice, 80
Wollaston, William, 153
wooden spoon, *defined*, 15
Woodhouse, Robert, 17, 22, 23
World's Columbian Exposition, 87
Worsley, Thomas, 43
wranglers, *defined*, 15
Wright, John, 18

Young, John Radford, 27
Young, Thomas, 151, 161